BRINGING SCANNING PROBE

MICROSCOPY UP TO SPEED

MICROSYSTEMS

Series Editor
Stephen D. Senturia
Massachusetts Institute of Technology

Editorial Board

Roger T. Howe, *University of California, Berkeley*
D. Jed Harrison, *University of Alberta*
Hiroyuki Fujita, *University of Tokyo*
Jan-Åke Schweitz, *Uppsala University*

Books in the Series

METHODOLOGY FOR THE MODELING AND SIMULATION OF MICROSYSTEMS
B. Romanowicz
ISBN: 0-7923-8306-0

MICROCANTILEVERS FOR ATOMIC FORCE MICROSCOPE DATA STORAGE
B.W. Chui
ISBN: 0-7923-8358-3

BRINGING SCANNING PROBE MICROSCOPY UP TO SPEED
S.C. Minne, S.R. Manalis, C.F. Quate
ISBN: 0-7923-8466-0

BRINGING SCANNING PROBE

MICROSCOPY UP TO SPEED

by

S. C. Minne
Nanodevices, Inc.

S. R. Manalis
Massachusetts Institute of Technology

C. F. Quate
Stanford University

KLUWER ACADEMIC PUBLISHERS
Boston / Dordrecht / London

Distributors for North, Central and South America:
Kluwer Academic Publishers
101 Philip Drive
Assinippi Park
Norwell, Massachusetts 02061 USA
Telephone (781) 871-6600
Fax (781) 871-6528
E-Mail <kluwer@wkap.com>

Distributors for all other countries:
Kluwer Academic Publishers Group
Distribution Centre
Post Office Box 322
3300 AH Dordrecht, THE NETHERLANDS
Telephone 31 78 6392 392
Fax 31 78 6546 474
E-Mail <orderdept@wkap.nl>

 Electronic Services <http://www.wkap.nl>

Library of Congress Cataloging-in-Publication Data

A C.I.P. Catalogue record for this book is available
from the Library of Congress.

BRINGING SCANNING PROBE

MICROSCOPY UP TO SPEED

Table of Contents

Introduction

The invention of the scanning tunneling microscope in 1982 by Binnig and Rohrer opened up many new areas in physics and surface science. The tunneling microscope, for the first time, provided a way to measure surface undulations on the atomic scale. The scientific community quickly accepted this innovation, and from it many novel surface science techniques were developed. Although the scanning tunneling microscope revolutionized the field of surface science, it was limited to the imaging of conductive materials.

The atomic force microscope was invented to image both insulating and conducting materials. Since its introduction in 1986, there have been many advances in the technology of cantilever fabrication, deflection detection, and electronic control. An entire field known as scanning probe microscopy has emerged with these advances. Scanning probe microscopes (SPM) are now routinely used to resolve a wide range of surface properties on the nanometer scale including topography, friction, magnetic field, and electric field. This allows scanning probes to be used for applications as diverse as DNA imaging, to inspecting defects of semiconductors, to measuring fundamental physical and chemical properties of surfaces.

However, the SPM generally takes several minutes to produce an image because the scanning speed is very slow. In addition to long imaging times, the maximum imaging area is limited to a few hundredths of a square millimeter. As a result, the widespread use of the SPM has been limited, and many areas of surface science and technology are untouched. If the scan speed and imaging area are increased, the superior resolution and versatility of scanning probes will allow the SPM to replace the scanning electron microscope for many applications.

In this book we will introduce the principles of scanning probe systems with particular emphasis of techniques for increasing speed. We will include useful information on the characteristics and limitations of current state-of-the-art machines as well as the properties of the systems that will follow in the future. Our basic approach is two fold. First we treat fast scanning systems for single probes and, second, we treat systems with multiple probes operating in parallel.

The key components of the SPM are the mechanical microcantilever with integrated tip and the system used to measure its deflection. In essence, the entire apparatus is devoted to moving the tip over a surface with a well-controlled force. The mechanical response of the actuator that governs the force is of utmost importance since it determines the scanning speed. The mechanical response relates directly to the size of the actuator: smaller is faster. Traditional scanning probe microscopes rely on piezoelectric tubes of centimeter size to move the probe. In future scanning probe systems the large actuators will be replaced with cantilevers where the actuators are integrated on the beam. These will be combined in arrays of multiple cantilevers with MEMS as the key technology for the fabrication process.

We will focus on the design of advanced cantilevers and cantilever arrays. We will introduce these topics by describing microcantilevers that use integrated piezoresistors for sensing the cantilever deflection. We will discuss the design of cantilevers with the addition of integrated actuators. We will show how these technologies lead to an increase in the speed of scanning probes. We then discuss the design and operation of cantilevers with optical deflection sensors that are suitable for arrays. A later chapter is devoted to a discussion of system with a magnetic tip suitable for application in magnetic systems for storage of digital data. It is an application that goes beyond imaging. The final chapters outline the MEMS process flow used for fabricating the cantilevers.

The book incorporates the work of the authors and several graduate students at Stanford University including Goksenin Yaralioglu, Jesse Adams, Hyongsok (Tom) Soh, and Kathryn Wilder. Professor Abdullah Atalar of Bilkent University in Turkey has been a constant and valuable collaborator for all the work described here. Jim Zesch and Babur Hadimioglu at Xerox provided crucial information on the zinc oxide films. Marco Tortonese, who introduced the piezoresistive cantilever, guided us with many valuable discussions, as did Virgil Elings of Digital Instruments, Inc. Profs. B. T. Khuri-Yakub and R. F. Pease at Stanford University guided us during microfabrication process development. Prof. A. Pisano of U. C. Berkeley and DARPA provided incisive comments and critical guidance. The Karel Urbanek Endowment, and Mr. William C. and Ms. Barbara H. Edwards, through the Leland T. Edwards fellowship, provided generous support for S. C. M. and S. R. M.

The authors are grateful to Dr. Nick Ulman for technical and strategic advice, writing several sections, and editing the book as a whole. Financial support for this work was provided by the Defense Advanced Research Projects Agency, the Office of Naval Research, the National Science Foundation, and the National Nanofabrication Facility and Department of Electrical Engineering at Stanford University.

S. C. Minne, S. R. Manalis, C. F. Quate

Palo Alto, Calif. 1999

CHAPTER 1

Improving Conventional Scanning Probe Microscopes

1.1 Introduction

The Scanning Probe Microscope (SPM), introduced in 1986, is now a common tool in many sectors of science and technology. It is a marvelous instrument with characteristics that are unsurpassed in many areas. Yet there is more to be done and we use this book to outline our vision for future instruments. We begin with a description of the essential components and emphasize those features which limit the performance and reduce the competitive edge of scanning probes.

The primary elements of the SPM are depicted in Figure 1.1.1. The image is formed by monitoring the deflection of the cantilever as the sample is scanned beneath the tip. The optical lever detection scheme shown in Figure 1.1.1 is commonly used in the commercial instruments due to its simplicity and sub-Angstrom sensitivity. The sample under observation is mounted on a PZT cylinder, known as the piezotube scanner. This cylinder is poled with a radial E-field which converts it to a piezoelectric material. The inner and outer surfaces of the cylindrical scanner are coated metal films which serve as contacts for the applied field. The outer conductor is segmented into four quadrants. With this arrangement the tube will bend in the lateral plane when a positive voltage is applied to the first quadrant and a negative voltage to the third quadrant. The tube is extended in the z direction (normal to the sample surface) with a uniform voltage applied to the four segments. The end of the tube with the sample attached moves in an arc but this is easily corrected by extending the tube at either end

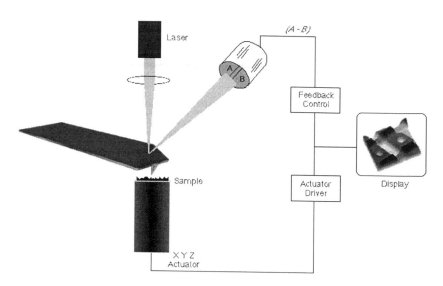

Figure 1.1.1   The essential features of the Scanning Probe Microscope

of the travel. The piezotube scanner is universally used as the scanning device in commercial instruments. However, this solution is not ideal. The piezotube scanner suffers from two limitations. The mechanical resonance is low (less than 1 kHz) and the bending angle is small. The speed of scanning is limited by the mechanical resonance and the image size is limited by the small bending angle. We believe that these instruments will be more widely used when the limitations are removed.

Microcantilevers with integrated tips (Figure 1.1.2) distinguish the SPM from all other microscopes. These devices are fabricated with MEMS technology using the processes outlined in Chapter 8. The micromachined cantilevers exhibit low spring constants for registering small forces (10^{-2} to 10^2 N/m) and high resonant frequencies (10 kHz to over 1 MHz) for fast data rates. The radius of curvature at the apex of the tip, with a typical value of 10 nm, sets the lateral resolution. Silicon and silicon nitride are the commonly used materials for the microcantilevers.

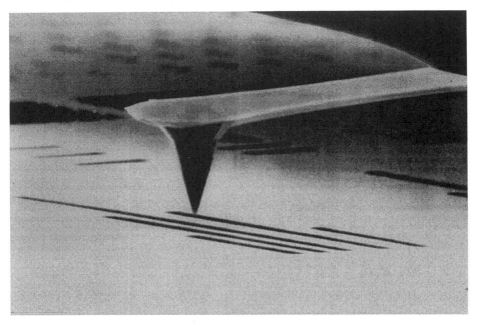

Figure 1.1.2   The microcantilever with integrated tip.   Photo courtesy IBM.

1.2 The Piezoresistive Cantilever

A piezoresistive cantilever incorporates a deflection sensor in the cantilever itself rather than using external sensing apparatus. The deflection sensor is based on the piezoresistive property of silicon. Piezoresistive cantilevers are well suited for closely spaced cantilever arrays where an external optical detection system would be impractical. Silicon cantilevers incorporating piezoresistive deflection detection were first demonstrated in 1991.[1]

---

1. M. Tortonese, H. Yamada, R. C. Barrett, and C. F. Quate, *Proceedeings of Transducers '91* (IEEE, Pennington, NJ, 1991), Publication No. 91 CH2817-5, p. 448.

We will find in a later chapter that the piezoresistive cantilever is not the ideal solution when an actuator is fabricated on the same cantilever. Nonetheless, it has been the workhorse and has played a strong role in most of the work on arrays.

This chapter presents the main features of piezoresistive cantilevers as a starting point to discuss more advanced cantilever designs. Subsequent chapters are concerned with developments that have occurred since the invention of the piezoresistive cantilever. We include some examples of what can be done with piezoresistive cantilvers by themselves and point out some of the limitations of these systems as motivation for the development of recent techniques.

The piezoresistive effect is described by:

$$\frac{\Delta\rho}{\rho}(x, y, z) = \pi_L S(x, y, z) \tag{1.2.1}$$

which says that the change in resistivity, ρ , of a material is proportional to the induced stress, S. The constant of proportionality is called the piezoresistive coefficient, π_L .

Stress in a cantilever is related to the force, P, at the tip by:

$$S(x, y, z) = \frac{P(x - L)y}{I} \tag{1.2.2}$$

where L is the length of the cantilever. I is the bending moment of inertia of the cantilever cross section and y is the distance from the plane of zero stress.

For the cantilever geometry shown in Figure 1.2.1, key parameters are spring constant, fractional change in resistance per unit displacement, and the minimum detectable deflection (MDD) . The spring constant, k, is the stiffness of the cantilever; i.e. the amount of force it takes to deflect the cantilever through a unit distance. For a regular, rectangular cantilever beam the spring constant is proportional to the cube of the cantilever thickness and inversely proportional to the cube of its length. The cantilver width and elasticity contribute linearly.

The fractional change in resistance, $\Delta R / R$, is a measure of the sensitivity of the cantilever to a displacement at the tip. Minimum detectable deflection (MDD) is defined as the deflection for a signal to noise ratio of unity.

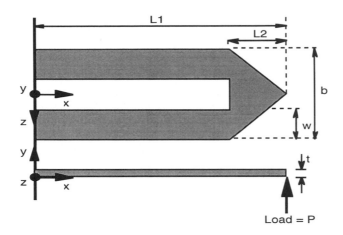

Figure 1.2.1   Cantilever coordinate system.

Changes in the piezoresistance are measured with a Wheatstone bridge biased to voltage, V. The output voltage is:

$$V_o = \frac{V}{4}\frac{\Delta R}{R}$$

(1.2.3)

The system is designed such that the dominant source of noise is the thermal noise of the piezoresistor. At higher cantilever biases, electrical heating introduces 1/f noise, causing an increase in the MDD. A cantilever bias of approximately 5V balances sensitivity and noise to give the small voltage optimum MDD. At 5V the dominant noise in the 0.01 Hz to 1 kHz bandwidth is 1/f noise, rather than thermal noise. At frequencies above about 100 Hz, the dominant noise source is thermal.

The parameters for a typical device[1] are, k = 16 N/m, $\Delta R/R$ = 0.9 ppm per Å, and MDD = 0.2Å. Theoretical and experimental values for k and $\Delta R/R$ agree closely. The minimum detectable deflection depends on the overall noise of the system.

These basic facts provide insight into the operation of the piezoresistive cantilever.

1. (L1=170 μm, L2=70 μm, w=17 μm, b=65 μm, and t=3.8 μm)

1.3 Imaging with Parallel Cantilevers

In our first attempt to increase the imaged area we fabricated parallel piezoresistive cantilevers on a common base. The linear array was used to demonstrate parallel imaging even though it was not possible to control the individual elements in the array. This experiment served to emphasize the need for individual actuators (and sensors) on each element in the array.

The cantilever die is fastened to an IC chip package and the contact pads of the piezoresistors are wire bonded to the output pins. The cantilever assembly is inserted into a lab-constructed AFM with the tips at a 15 degree angle to the sample. To convert changes in cantilever resistance into changes in voltage, the leads from the

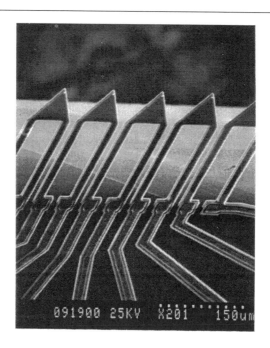

*Figure 1.3.1   SEM micrograph of an array of five
piezoresistive cantilevers.*

piezoresistors are connected to individual Wheatstone bridges biased at 4 V.

During tip-to-sample approach, the output signals from the cantilevers positioned at the ends of the array are monitored as the cantilevers are brought into contact with the sample. When one of the end cantilevers makes contact with the sample, the approach is stopped, and the angle of the array is adjusted. After levelling, all the tips are brought into contact with the surface and each cantilever is operated as a standard AFM.

To create an image, we multiplex the signals from the cantilever array. Each individual cantilever is selected from a single output line according to its binary address. This configuration reduces the number bridge connections for a parallel array of N cantilevers to $\log_2 N$. In Figure 1.3.2 we show a constant height image of a 400 μm x

Figure 1.3.2   400 μm x 100 μm parallel,constant height image. The dimensions of
"STANFORDQUATE" are 400 μm x 37 μm x 5000
on silicon.

100 μm area. It is recorded with four cantilevers operating in parallel. The letters in this image were formed with silicon dioxide on a silicon substrate. The spring constant of the cantilevers used in this image is 4 N/m, and the minimum detectable deflection is 0.4 Å in a 100 Hz to 1 kHz bandwidth. The probes were arranged to increase the image area for a given scan size and scan time. Alternatively, we can arrange the probes to decrease the scan time while holding the image area constant. Thus, parallel imaging provides a mechanism where either the acquisition time is decreased or the scan size is increased by a factor equal to the number of probes acting in parallel.

With this array, the images are taken in the constant height mode. Since there is only one z-axis control for the entire array of cantilevers, we can not individually control the force on each cantilever. It is possible to use feedback control for one of the cantilevers in the array, but since each tip experiences a different interaction with the sample, this is not ideal. For optimum imaging applications we require integrated actuators on each cantilever as outlined in the next chapter.

CHAPTER 2

Design of Piezoresistive Cantilevers with Integrated Actuators

2.1 Introduction

Simultaneous imaging with an array of cantilevers requires z-axis control of each element of the array. This means that we must replace the bulk actuator with an actuator designed in such a way that it can be integrated onto the individual cantilevers.

In this chapter we will discuss the design of the new actuator. It is a piezoelectric film of zinc oxide (ZnO) deposited onto the cantilever beam. This integrated piezoelectric actuator together with the piezoresistive sensor provides a modular cantilever design that is incorporated into each element of the array.

This experience has taught us that those instruments with just a single cantilever are improved with the addition of the integrated actuator. The piezotube used in standard contact-mode scanning probes limits the z-axis mechanical response. Its entire mass must be moved to change the height of the tip as it scans across the sample topography. In the next chapter we show that the scanning speed of a typical microscope is improved by more than one order of magnitude with the integrated actuator.

In our device the actuator is created as a bimorph with a film of ZnO on the cantilever. When a voltage is applied across the ZnO, causing it to expand or contract, the tip will move along the z-axis and change the tip-to-sample spacing.

Since minimum detectable deflection (MDD) is the parameter to be optimized, we will see that the ZnO region should be made thick in order to increase its bending moment of inertia. However, there is a point of diminishing returns. A ZnO layer that is too thick will reduce the deflection and increase coupling between the actuator and sensor because the plane of zero stress is moved too far away from the surface of the silicon

A reasonable compromise for these trade-offs can be found in the geometry that we present in this chapter. We refer to this as the "standard geometry", specified in Table 1, on page 29.

There are basic requirements for any cantilever that must be considered when a new design is attempted. First, the actuator must not induce a sensor signal when the tip is not touching the sample. In other words the system must be able to reproduce the standard force versus distance curve which is flat when the tip is not in contact, and steadily rising after contact is made. Second, to replicate the operation of a standard AFM, the cantilever must have separate sensor and actuator regions; the sensor cannot be in the same place as the actuator. The actuator has to be able to move the base end (away from the tip) of the sensor region of the cantilever.

Mathematical analysis of the ZnO cantilever is complicated due to the number of different layers in the structure and the nonuniform geometry along its length. In the next several sections we will discuss this combination with both closed form and numerical analyses. Comparison of piezoresistive cantilevers with and without actuators is useful in understanding of how the various layers affect cantilever performance. Given here for reference are the basic formulas for spring constant, sensitivity, and minimum detectable deflection for a cantilever without an integrated actuator:

$$k = \frac{P}{\delta(L1)} = \frac{Yt^3wb}{2b(L_1^3 - L_2^3) + 6wL_2^3} \tag{2.1.1}$$

$$\left(\frac{\Delta R}{R}\right) = \frac{3\pi_L Ytb(L_1 + L_2)}{4[(L_1^3 - L_2^3)b + 3L_2^3w]}y \tag{2.1.2}$$

$$MDD = \frac{16[(L_1^3 - L_2^3)b + 3L_2^3w]}{3\pi_L Ytb(L_1 + L_2)}\frac{\sqrt{2k_bTR\Delta f}}{V} \tag{2.1.3}$$

where π_L is the piezoresistive coefficient, and $\sqrt{4k_bTR\Delta f}$ is the thermal resistor noise and the other variables are defined in Figure 1.2.1 on page 19.

2.2 The ZnO / Piezoresistive Cantilever

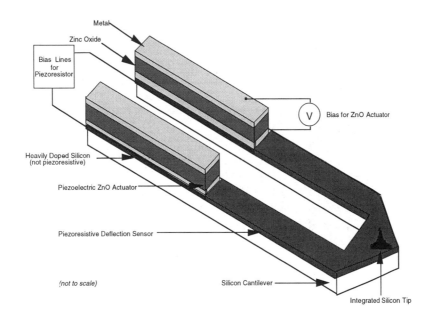

Figure 2.2.1   Schematic diagram of a ZnO / piezoresistive cantilever.

A schematic diagram of a cantilever containing an integrated piezoelectric actuator and piezoresistive sensor is shown in Figure 2.2.1. An SEM micrograph of an array of two cantilevers is shown in Figure 2.2.2. Each cantilever is 420 μm in length with the ZnO base region occupying 180 μm of the total length. The full width of each cantilever is 85 μm, each leg of the cantilever is 37 μm in width, and the tips are spaced 100 μm apart. The actuator is constructed by patterning a thin film of ZnO, onto the base of a single crystal cantilever. The piezoresistive sensor is formed by lightly doping the remaining portion of the cantilever with an ion implant. To insure that the ZnO does not mechanically couple to the piezoresistor when actuated, the sil-

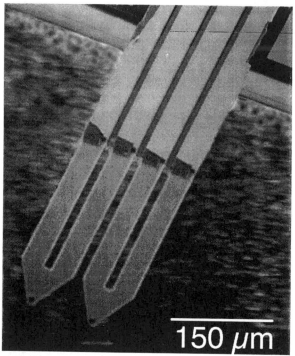

Figure 2.2.2   SEM micrograph of two piezoresistive cantilevers.

icon beneath the ZnO is heavily doped. Metal electrodes are patterned above and below the ZnO such that a voltage can be applied across the film. These electrodes, along with contacts to the piezoresistor, are connected to large bond pads located on the cantilever die. Wire bonds are used to connect the pads to a standard DIP package.

The design of the silicon beam is governed by several factors such as the resonant frequency (20 kHz to 70 kHz) and the spring constant (0.6 to 7.1 N/m). The silicon beam must be thick enough to accommodate the piezoresistor. The piezoelectric film thickness is chosen to maximize the deflection of the silicon beam resulting from an applied voltage across the ZnO for a given cantilever design. For thin ZnO films, the bending force will increase with film thickness, but there is a limit. When the ZnO layer is thick compared to the silicon, the cantilever is merely a slab of ZnO. In this case the applied voltage elongates the beam; it does not bend it. For bending, the ZnO film must be confined to the upper half of the cantilever. In our design the optimum thickness of the ZnO film is roughly equal to the thickness of the silicon beam, or 3.5 μm.

The cantilever bends from an applied potential to the ZnO actuator due to the bimorph effect as depicted in Figure 2.2.3 The ZnO is oriented such that when an electric field

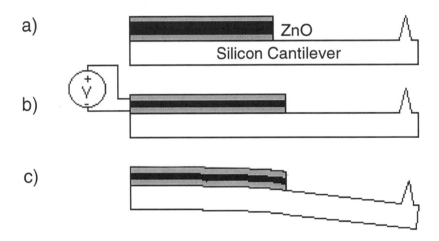

Figure 2.2.3   Silicon cantilever bending under stress applied by a ZnO microactuator.

is applied across the electrodes, the piezoelectric properties of the film cause the length to either expand or contract depending on the electric field orientation. This change creates stress in the ZnO film which must be relieved through bending of the silicon cantilever beneath the ZnO. Equilibrium is obtained once the bimorph region of the cantilever is deformed and this causes the tip to be displaced vertically. The maximum tip displacement for a voltage swing of ±35 V is roughly 4 μm for a cantilever that is 720 μm long. Higher voltages cause electrical breakdown of the ZnO.

A key feature of the ZnO cantilever is that the bimorph portion is nearly an order of magnitude stiffer than the piezoresistive region. The large difference in the lever stiffness occurs because the cantilever spring constant is proportional to the thickness cubed and the ZnO region is twice as thick as the piezoresistive region. As a result, forces applied by the sample topography primarily bend the flexible piezoresistive region. This feature allows the cantilever to sense and control the tip/sample force simultaneously.

2.3 Theory of Operation

Microcantilevers for scanning probe systems are characterized by four important per-

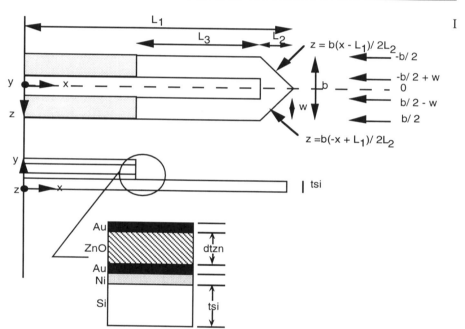

Figure 2.3.1   ZnO cantilever geometry.

formance parameters: spring constant, maximum induced deflection, minimum detectable deflection, and actuator-sensor coupling. In the rest of this chapter we show how to calculate these parameters using the coordinate system shown in Figure 2.3.1. In many of the calculations numerical values will be inserted to provide an intuitive feel. When this is done, the values will be those of the "standard geometry" in TABLE 1.

L_1	$7 L_3 / 3 = 420 \ \mu m$
L_2	$L_3 / 3 = 60 \ \mu m$
L_3	$L_3 = 180 \ \mu m$
b	$85 \ \mu m$
w	$(b - 10) / 2 = 37 \ \mu m$
Gold Electrode Thicknesses	$0.5 \ \mu m$
Nitride Buffer Thickness	$0.2 \ \mu m$
ZnO Thickness	$3.5 \ \mu m$
Cantilever Thickness	$3.5 \ \mu m$
Youngs Modulus of Silicon	$1.7 \ x \ 10^{11} \ N/m^2$
Piezoelectric Coefficient ZnO	$d_{31} = 5.4 \ x \ 10^{-12} \ C/N$
Piezoresistive Coefficient	High Doping,$1.1 \ x \ 10^{-10} \ m^2/N$
	Low Doping, $5.7 \ x \ 10^{-10} \ m^2/N$

TABLE 1. The "Standard Geometry" ZnO Cantilever

The spring constant, k, is the stiffness of the cantilever. Current commercially available cantilevers have spring constants ranging from tenths to hundreds of Newtons per meter. In general, cantilevers with low spring constants are desirable for contact-mode imaging because they apply less force to a sample for a given deflection.

The maximum induced deflection of a cantilever is how far the actuator can move the cantilever from its relaxed position. For leveling, imaging, and vibration applications, a one micron induced deflection is generally sufficient.

The minimum detectable deflection (MDD) determines the sensitivity of the cantilever. In today's commercial systems this value is generally a few tenths of an Angstrom. MDD is calculated by finding the cantilever deflection that produces an output voltage equal to the rms noise voltage of the system. For a piezoresistive system, MDD is inversely proportional to $\Delta R / R$.

When an actuator is integrated in the cantilever, an electrical signal to the actuator moves the tip. This allows feedback imaging, provided the actuator does not create a spurious signal in the sensor when activated. When the tip is not in contact, the amount of deflection that the piezoresistor senses for a given deflection of the tip by the actuator is the actuator-sensor coupling. Ideally this value should be zero.

Before any of the performance parameters can be calculated, however, we need to find the centroid and moment of inertia.

2.4 Centroid and Moment of Inertia

The first task in analyzing a cantilever is to find its centroid and moment of inertia. The centroid is the neutral bending axis, or the plane within the cantilever that does not experience any stress when the cantilever is bent.

Direct mechanical analysis of the multi-layer structure is most simply done using the theory of equivalent widths. The theory says: "A multi-layer cross-section can be converted into an equivalent cross-section of any reference material; provided the stresses in each material remain within the proportional limit for that material. Each material's width is adjusted about the principle axis of bending by the ratio of the Modulus of Elasticity of that material to the reference material."[1]

The cross sections through our cantilever geometry are shown in Figure 2.4.1.

Young's moduli for Au, Ni_3Si_4, ZnO, and Si are 0.8, 1.4, 1.2, and 1.7 x10^{11} N/cm^2 respectively. Converting to an equivalent profile of silicon, the new widths become: 0.47w for Au, 0.82w for Ni_3Si_4, and 0.71w for ZnO. The equivalent silicon profile in the ZnO region is shown below. The other regions consist of only silicon, so their profiles do not change.

Since the centroid is involved in future calculations, it must be known. In addition to computational utility, the centroid's location can provide an intuitive feel for cantilever performance. The piezoresistor is located on the top surface of the silicon at position t_{si}. The centroid is the plane of zero stress, so we can estimate the piezoresistor's relative strength according to its distance from the centroid. In order to optimize sensitivity, we want to design the cantilever such that the piezoresistor is as far from the centroid as possible. There are some exceptions to this rule in the ZnO region.

In our design, the actuator uses the bimorph effect to move the tip vertically. This movement creates stress within the silicon cantilever, and thus the piezoresistor. The

---

1. S. Timoshenko, "Strength of Materials," 3rd ed., New Delhi Affiliated East-West Press, Princeton, N. J. (1958).

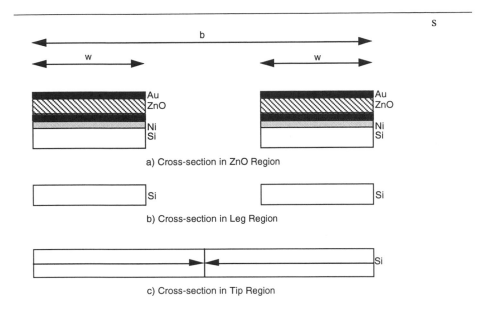

S

a) Cross-section in ZnO Region

b) Cross-section in Leg Region

c) Cross-section in Tip Region

Figure 2.4.1   Cross sections of the cantilever geometry.

amount of piezoresistive signal that the ZnO actuator induces is unwanted coupling. Therefore in this region, the piezoresistor should be as close as possible to the zero stress plane, or centroid.

The position of the centroid must also be considered in the mechanical design. When the ZnO is activated it will produce a uniform internal stress. If the ZnO is on one side of the centroid it will produce bending as expected. However, if the ZnO straddles the centroid, the induced stress will not only provide bending, but it will cause elongation as well. For a very thick ZnO region, the centroid will approach the center of the ZnO film, causing any induced stress to be equal on each side of the neutral axis. At that point, the ZnO activation will cause pure elongation and no bending. Therefore the position of the centroid relative to the ZnO will determine the bending efficiency of the ZnO actuator.

We are only concerned with bending in a single direction, so we need only find the vertical centroid. Centroid is defined as:

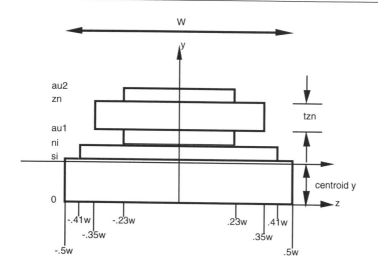

Figure 2.4.2   Equivalent silicon profile of the ZnO region.

$$\bar{y} = \frac{1}{Area} \int y \, dA \qquad\qquad (2.4.1)$$

and has units of μm in our calculations. Notice that when we defined the coordinate system we normalized all widths to the width of the silicon, therefore the width cancels from (2.4.1).

For the leg and tip regions, the centroid is simply located half way through the silicon, or at $t_{si}/2$. Due to the large number of layers in the ZnO region, the symbolic centroid equation is quite complicated. The actual equations are best handled by numerical analysis programs such as Mathematica. The centroid value can be found from Figure 2.4.3 for different silicon and ZnO thicknesses after applying the standard geometry to the gold and nitride layers.

The moment of inertia is another geometric constant that is necessary for calculation, and also serves as a valuable intuitive guide. The moment of inertia (MOI) tells us how much a given geometry will deflect for an applied moment, or force. Within our structure we have three regions, each with a different MOI. Comparison is valuable because it will tell how much each region will deflect for a load at the tip, and when actuating the ZnO, how easily the tip region will be influenced. MOI also determines the amount of stress generated in a beam for a given bending moment. Since the

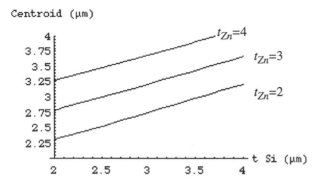

Figure 2.4.3 Location of the centroid in the ZnO region for various silicon and ZnO thicknesses.

piezoresistor measures stress, this is an important consideration. For the aspects of the cantilever performance listed above, we desire a large MOI in the ZnO region and a small MOI in the piezoresistive region. This insures that we have a stiff actuator bending a flexible sensor.

Since MOI is the second moment of the cross section, the thick ZnO base provides a desirable MOI distribution through the cantilever. The equation for MOI is:

$$I = \int y^2 dA \qquad (2.4.2)$$

and has units of μm^4 in our calculations.

For a rectangular geometry, like that of the piezoresistive region,

$$I = \frac{2wt_{Si}^3}{12} \qquad (2.4.3)$$

where the factor of 2 is to account for both legs of the cantilever. To calculate MOI in the tip region the variation of width versus position is substituted for the (2w) in equation (2.4.3). In the ZnO region, the equation is more complicated, and is best examined numerically. Moment of inertia is plotted along the cantilever length for the standard geometry. As expected from the cubic dependence of thickness on MOI, for

---

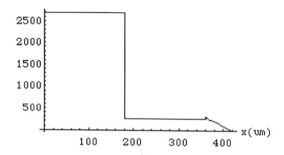

Figure 2.4.4   *Moment of inertia throughout the length of the cantilever. Cantilever length is 420 um, 3/7 of which is the ZnO region, 3/7 piezoresistive region, and 1/7 tip region.*

this geometry, the moment of inertia in the ZnO region is roughly an order of magnitude greater than in the piezoresistive region.

2.5 Spring Constant

The equation for spring constant, k, is:

$$k = \frac{F(L_1)}{y(L_1)}$$

(2.5.1)

where $F(L_1)$ is force applied to the cantilever tip, and $y(L_1)$ is the vertical displacement at the tip. In the following calculations the distributed load (ω) is modeled by a delta function of magnitude P located at the tip, $\omega(x) = P \delta(x-L_1)$. The displacement, y, can be found by applying the following equations:

$$V = \int \omega dx + C_1$$

(2.5.2)

$$M = \int V dx + C_2$$

(2.5.3)

$$\theta = \int \frac{M}{YI} dx + C_3 \qquad\qquad (2.5.4)$$

$$\delta = \int \theta dx + C_4 \qquad\qquad (2.5.5)$$

where V is shear, M is the bending moment, θ is the bend angle, δ is deflection, Y is Young's modulus, I is the moment of inertia, and the constants (C1, C2, C3, C4) are evaluated according to the boundary conditions.

Since the load is simply a delta function at the tip, the shear is a constant and the bending moment is the line $M = P (x - L1)$. The inverse of radius of curvature is found by:

$$\frac{1}{\rho} = \frac{M}{IY} \qquad\qquad (2.5.6)$$

$1/\rho$ is integrated to find slope, which is integrated to find deflection. Figure 2.5.1 plots slope and deflection through a cantilever with the standard geometry.

From these plots it can be seen that when a force is applied at the tip most of the bending occurs in the piezoresistive region. The cantilever displacement at the end of the ZnO region is 0.012 µm, or less than 10% of the displacement at the tip (0.147 µm). The spring constant is the load divided by the deflection or 6.8 N/m.

Since the symbolic result for the standard geometry is quite complex, insight into the spring constant calculation can be found by considering bending in a simple beam. It turns out that in our device, spring constant is a factor in both sensitivity and actuator-sensor coupling. Section 2.7 will show that the reduction in spring constant from the addition of the actuator proportionally decreases the sensitivity of the piezoresistor.

Analysis of a simple cantilever beam of length L, an MOI of I, and an applied force at the tip of P produces a deflection and spring constant of:

$$\delta = \frac{PL^3}{3YI} \qquad\qquad (2.5.7)$$

$$k_{simple} = \frac{P}{\delta} = \frac{3YI}{L^3} \qquad\qquad (2.5.8)$$

Design of Piezoresistive Cantilevers with Integrated Actuators   **35**

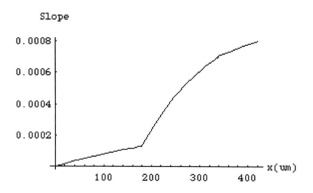

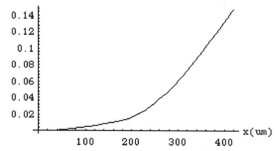

Figure 2.5.1   Slope and deflection of the ZnO cantilever for a 1 μN load at the tip.

If the cantilever is extended in length by adding a base of length (L_1 - L), and a MOI of I_{big}, as shown if Figure 2.5.2 , the new deflection for the same applied force is:

$$\delta = \frac{P}{Y}\left(\frac{L^3 - L_1^{\,3}}{3I_{big}} - \frac{L^3}{3I}\right)$$

(2.5.9)

If we assume that L, the original thin region, is some fraction of the total length L_1, or $L = nL_1$ where n can vary from 0 to 1, then the spring constant for the compound cantilever becomes:

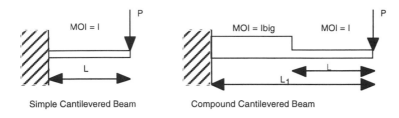

Figure 2.5.2   Simple and compound cantilever beams.

$$k_{compound} = \frac{3YI}{(nL_1)^3}\left[\frac{1}{1 + \frac{I}{I_{big}}\left(\frac{1}{n^3} - 1\right)}\right] = k_{simple} \cdot \left[\frac{1}{1 + \frac{I}{I_{big}}\left(\frac{1}{n^3} - 1\right)}\right] \qquad \textbf{(2.5.10)}$$

or the original thin beam spring constant adjusted by a correction factor or "modifier".

To apply equation (2.5.2) on page 34 to our standard geometry, we first need to find n, or the fraction of the cantilever without the actuator. Grouping the leg region and the tip region into the thin section, and ZnO region as the thick section, n is 4/7. From Figure 2.4.4 the ratio I_{thick} to I_{thin} in the standard geometry is about 10. Applying these two values to equation (2.5.2) the spring constant modifier for adding an actuator to a piezoresistive cantilever is 0.69. Complete numerical simulation of the standard geometry, with and without an actuator region, produces a modifier value of 0.70. The close agreement is expected as the only simplification made is the exclusion of the effects of the tapered tip region on MOI.

We can now examine the effect of ZnO thickness, through MOI, on spring constant. Figure 2.5.3a shows the dependence of the spring constant modifier on the ratio of MOI for n = 4/7. Increasing the MOI ratio to values larger than 10 has diminishing returns for improving the spring constant modifier and sensitivity. Even so, if the ratio is increased, or if the ZnO thickness is increased, maximum induced deflection will suffer (see section 2.6) due to relocation of the centroid, and fabrication complexity will increase. Figure 2.5.3b shows examples of spring constants for the standard geometry for different cantilever lengths.

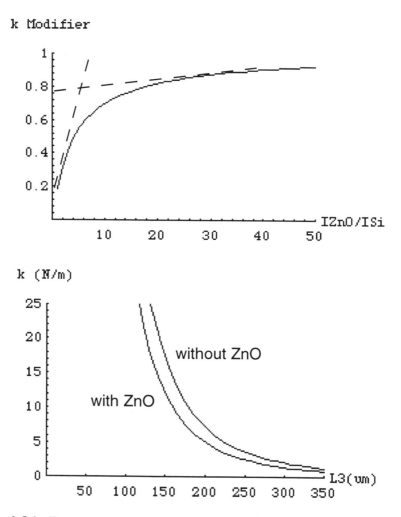

Figure 2.5.3   Top: variation in spring constant scale factor vs. MOI ratio. Bottom: Spring constant vs. length L_3.

In equation (2.5.2), for our geometry, n represents the fraction of the cantilever that does not contain the actuator. As expected, varying n causes the spring constant to vary between the two cases of beams with MOI of I and Ibig, and a length of L_1. The implication of changes in n is not as intuitive because of the cubic dependence of n in both the spring constant and the modifier. In practice, a specific ZnO length is needed for a given actuation specification, and n is then tailored to meet stiffness or sensitivity requirements.

2.6 Maximum Induced Deflection

Applying a voltage to the ZnO causes it to contract or expand. That expansion creates a stress which, for equilibrium, must be compensated by an equal and opposite stress in the rest of cantilever structure. Returning to Figure 2.4.2 on page 32 we see that the ZnO is isolated on one side of the centroid, therefore the stress opposing the ZnO must be produced on other side. In any structure, when there are opposite stresses on each side of the neutral axis, bending occurs.

For the compound cantilever the total deflection of the cantilever tip will be the sum of the bending due to deflection in the actuator region, plus the slope at the end of the actuator multiplied by the length of the unactuated region. Figure 2.6.1 shows a sche-

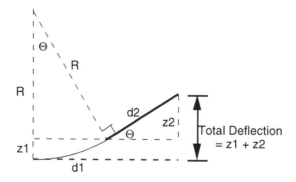

Figure 2.6.1   Bending of a compound cantilever.

matic of the deflection components. For small deflections, the total deflection in Figure 2.6.1 is given by:

$$Def = \frac{d_1^2}{2R} + \frac{d_1 d_2}{R}$$

(2.6.1)

or 11 $L_3^2/$ 6 R using the standard geometry. Equation (2.6.1) can either be found geometrically or by integrating the inverse radius of curvature twice over the cantilever length.[2]

In either case radius of curvature must be related to the applied electric field. The radius of curvature is found by calculating the value where the bending moment through a cross-section of the actuator is equal to the induced ZnO bending moment. Radius of curvature is related to bending moment by:

$$M = \frac{YI}{\rho}$$

(2.6.2)

where Y is Young's modulus, I is the moment of inertia, and M is the bending moment.

The ZnO induced bending moment is found by multiplying the piezoelectric strain (d_{31} E), Youngs Modulus (to convert to stress), area (to convert to force), and distance from the centroid (to convert to bending moment). In symbolic form,

$$M = d_{31}EY\int\int y \, dy \, dz$$

(2.6.3)

where d_{31} is the piezoelectric coefficient, and E is the electric field across the piezoelectric material. For ZnO the piezoelectric coefficient is 5.43 x 10^{-12} C/N. For equilibrium conditions the net bending moment equals zero:

$$M = 0 = \frac{YI}{\rho} + d_{31}EY_{ZnO}\int\int y \, dy \, dz$$

(2.6.4)

Solving for ρ:

$$\frac{1}{\rho} = \frac{d_{31}EY_{ZnO}\int\int y \, dy \, dz}{YI}$$

(2.6.5)

To find deflection at the tip substitute (2.6.5) into (2.6.1). By examining (2.6.1) and (2.6.5) we clearly see the important factors for maximizing the tip deflection. A long cantilever completely covered with ZnO would have the maximum amount of bend-

---

2. Note: $\frac{1}{\rho} = \left(\frac{d^2y}{dx^2}\right)/\left[1+\left(\frac{dy}{dx}\right)^2\right]^{3/2}$ which is approximated by $\frac{d^2y}{dx^2}$ for small dy/dx.

ing, however, the uncovered extension only creates a minor reduction in the induced deflection. Radius of curvature should be minimized. Electric field is used to control the deflection and it should be used up to the breakdown field of the ZnO, or about 10^7 V/m.

Determining how other parameters affect the deflection is more complicated because of the interdependence of the film thickness on the moment of inertia and the centroid. Simply making the ZnO thicker would increase the MOI causing $1/\rho$ to decrease, however it would also change the position of the centroid causing the integral to change. The net effect can not be easily seen from Equation (2.6.5). Figure 2.6.2 plots the ZnO induced tip deflection versus ZnO thickness for the "stan-

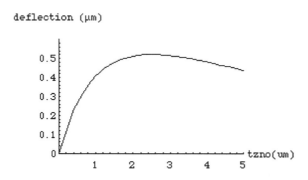

Figure 2.6.2   Tip deflection versus ZnO thickness

dard geometry". In the plot electric field is held constant, not voltage. If voltage is held constant bending will be maximum for zero ZnO thickness. Although this is mathematically true, there would be an infinite electric field in the ZnO. Breakdown of the film must be considered.

From Figure 2.6.2 we see that for thin layers of ZnO, the film is far from the neutral axis and will be very efficient at bending. Increasing the thickness of the film will increase its effectiveness. However, there is a limit. If the ZnO were infinitely thick, the cantilever would be just a slab of ZnO and an applied field would cause only elongation, not bending. Figure 2.6.2 demonstrates this graphically.

For fabrication, we used a ZnO thickness of 3.5 μm even though this thickness is not the theoretically optimum value for deflection. It is better to operate on the less steep side of the curve because of nonuniformities in the ZnO deposition and in the etches that define the cantilever thicknesses. Also, a thicker ZnO film will increase the MOI in the ZnO region, increasing the spring constant modifier which increases the piezoresistor sensitivity. Increasing the thickness of the ZnO by one micron greater than the optimum changes the tip deflection from 0.52 μm to 0.50 μm. Figure 2.6.3 plots a side view of the cantilever deflection for different ZnO biases.

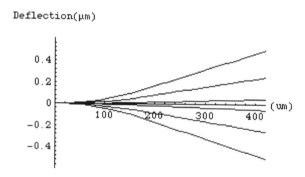

Figure 2.6.3   Deflection versus bias. From top to bottom the curves correspond to biases of 10, 5 ,1, -1, -5, and -10 x 10⁶ V/m on the ZnO region.

2.7 Minimum Detectable Deflection

The minimum detectable deflection is the deflection that causes the output voltage of the cantilever to be equal to the rms noise voltage. For a given cantilever operating point the MDD is inversely proportional to the piezoresistive effect, $\Delta R/R$. In a cantilever with two piezoresistive regions, $\Delta R/R$ is given by the following equation:

$$\frac{\Delta R}{R} = \frac{\dfrac{\Delta R_{ZnO\ Region}}{R_{ZnO\ Region}} R_{ZnO\ Region} + \dfrac{\Delta R_{Si\ Region}}{R_{Si\ Region}} R_{Si\ Region}}{R_{ZnO\ Region} + R_{Si\ Region}} \qquad \text{(2.7.1)}$$

For a piezoresistive element the relationship between normalized change in resistivity and stress is given by:

$$\frac{\Delta\rho}{\rho}(x, y, z) = \pi_L S(x, y, z) \qquad (2.7.2)$$

where ρ is resistivity, π_L is the piezoresistive coefficient (which is a function of doping)[3], and S is the stress. Stress in the cantilever is given by:

$$S = \frac{My}{I} \qquad (2.7.3)$$

where y is the distance from the centroid.

It is important to note that the piezoresistors in the ZnO region and the piezoresistor in the silicon region can be on opposite sides of the centroid. In other words, a deflection at the tip that induces compressive stress in one region, could cause tensile stress in the other region. The net effect would be some cancellation of the two piezoresistive effects, or an increase of the MDD.

Another important consideration about the piezoresistor in the ZnO region is actuator-sensor coupling. Section 2.6 described how the ZnO induces a stress in the silicon to create bending. Equation (2.7.2) shows that the stress will cause a change in resistance, or a deflection signal. Clearly we do not want our feedback actuator to create a spurious sensor signal.

For these two reasons it necessary to eliminate the piezoresistive contribution from the silicon beneath the ZnO. In our design this is accomplished in three ways. First, the ZnO is made slightly thicker than optimum in order to move the centroid to the surface of the silicon, reducing the stress in the piezoresistive region. Second, the doping density of the silicon under the ZnO is increased by a factor of 100 over the density in the piezoresistor silicon.[4] This causes the resistance[5] of the ZnO region to

---

3. Y. Kanda, "A Graphical Representation of the Piezoresistance Coefficients in Silicon," IEEE Trans. Ele. Dev., **ED-29**, 64 (1982)

4. Alternatively, metal leads could be run under the ZnO to carry the piezoresistor bias to the tip region. This would eliminate the piezoresistive coupling, but it would also significantly increase the fabricational complexity.

5. Resistance is found by: $R = \frac{\rho L}{wt} = \frac{L}{(q\mu_p p)wt}$ where q is the electron charge, μ is the mobility and p is the doping density.

be ~100 times less than that of the piezoresistor. As seen in equation (2.7.1), the lowered resistance reduces the ZnO region's contribution to the overall piezoresistive effect. Third, the piezoresistive coefficient π_L decreases with doping. The hundred fold increase in doping decreases π_L by 80%. This causes a direct reduction in the sensitivity of the piezoresistor beneath the ZnO. The actual actuator sensor coupling will be calculated in Section 2.8.

The total $\Delta R / R$ can be calculated by assuming the doping profiles are step functions, with values of 4×10^{18} cm^{-3} extending 0.5 microns into the silicon for the piezoresistive region, and 10^{20} cm^{-3} extending 0.9 microns into the silicon for the ZnO region. Applying the standard geometry of Table 1, on page 29, the mobility and resistance of the two regions are: $\mu_{ZnO \, region} = 50$ cm^2/Vs, $\mu_{Si \, region} = 100$ cm^2/Vs,[6] and $R_{ZnO} = 313 \, L_3 / w$, $R_{Si} = 13.9 \, L_3 / w$.

$\Delta R / R$ for the two regions is found by integrating and normalizing (2.7.2):

$$\frac{\Delta R}{R} = \frac{1}{L} \int \frac{\Delta \rho}{\rho} \, dx = \frac{1}{L} \int \pi_L \frac{My}{I} dx \qquad (2.7.4)$$

It is convenient to normalize $\Delta R / R$ to a unit deflection at the tip. Fortunately bending moment, M, is proportional to the load at the tip, P, which is related to the deflection at the tip through the spring constant, k in equation (2.5.1). Making this substitution into (2.7.4) we find:

$$\frac{\Delta R}{R} \text{per unit deflection} = \frac{1}{\Delta L} \int_{\Delta L} \pi_L \frac{k(x - L_1)y}{I} \, dx \qquad (2.7.5)$$

Applying the standard geometry:

$$\frac{\Delta R_{ZnO}}{R_{ZnO}} = -1.63 \times 10^{-9} \text{ per Å}, \quad \frac{\Delta R_{Si}}{R_{Si}} = 3.85 \times 10^{-7} \text{ per Å} \qquad (2.7.6)$$

and using $R_{ZnO} = 67 \, \Omega$ and $R_{Si} = 1.5$ kΩ, from (2.7.1) total $\Delta R / R$ is 3.68 x 10^{-7} per Å. The piezoresistive coefficient for the Si region is 5.7 x 10^{-10} m^2/N and (0.2) 5.7 x 10^{-10} m^2/N for the ZnO region. From heavy doping and the positioning of the centroid, the ZnO region contributes less than 1% to the overall $\Delta R / R$.

---

6. B. Streetman, Solid State Electronic Devices, 2nd Ed., Prentice-Hall Inc., Engelwood, N. J. (1980).

It is useful to compare how the addition of an actuator onto the cantilever structure affects the sensitivity of the device. Since the doping in the ZnO region is so high, $\Delta R/R$ and R in that region are negligible when compared to the leg region. With this assumption, the calculation is direct because the only variable in equation (2.7.5) that changes with the addition of the actuator is spring constant, k. Spring constant is determined by the initial geometry of the device and can be considered a constant in the integral. Taking the sensitivity ratio of two cantilevers of the same geometry, with and without actuators, the integrals cancel and we are left with only spring constant modifier (k_{ZnO}/k_{PR}). For the standard geometry we previously calculated this to be 70%.

2.8 Actuator-Sensor Coupling

As discussed earlier, we do not want to create a false deflection signal by coupling stress induced by the actuator into the piezoresistor. By carefully positioning the centroid and increasing the doping in the silicon under the ZnO region we minimize this coupling. The severity of coupling can be calculated by finding the ratio of $\Delta R/R$ per unit deflection induced by the ZnO and $\Delta R/R$ per unit deflection for a displacement at the tip. If this ratio is small, it means that the amount of stress that the actuator induces to correct for a tip displacement is negligible compared to the actual deflection signal. In the previous section we found $\Delta R/R$ for a deflection at the tip. $\Delta R/R$ induced by the ZnO is found in the following manner:

Equation (2.6.3) shows that the ZnO induced bending moment is constant throughout the length of the ZnO. From (2.7.2), (2.7.3), (2.7.4), and (2.6.2) we can calculate:

$$\frac{\Delta \rho}{\rho} = \frac{\Delta R}{R} = \pi_L S = \frac{\pi_L Y_{Si} y}{\rho} = \pi_L Y_{Si} y \, \Im(y) E \qquad (2.8.1)$$

where $\Im(y)E$ represents the right side of equation (2.6.5). To find the deflection at the tip for a given ZnO bias we again use equation (2.6.5), but this time we integrate twice. (It must be integrated over the entire length of the cantilever. In the leg region, $1/\rho = 0$, since there is no actuator) The result is:

$$\delta(L_1) = \frac{(-L_1 + L_2 + L_3)(L_1 + L_2 + L_3)}{2} \, \Im(y) E \qquad (2.8.2)$$

Dividing equation (2.8.1) by (2.8.2) $\Delta R/R$ per unit deflection is found:

Design of Piezoresistive Cantilevers with Integrated Actuators **45**

$$\frac{\Delta R}{R} \text{per Å} = \frac{2\pi_L Y_{Si} y}{(-L_1 + L_2 + L_3)(L_1 + L_2 + L_3)} \tag{2.8.3}$$

The importance of minimizing the piezoresistive coefficient (π_L) and the distance from the centroid (y) are evident from equation (2.8.3). Applying the standard geometry the actuator's contribution to $\Delta R/R$ is 5.66 x 10^{-9} per Å. $\Delta R/R$ for the cantilever was 3.68 x 10^{-7} per Å, so for the standard geometry the coupling between the actuator and the sensor is less than 2%.

2.9 Comparison to other Analyses

There are many papers that analyze a piezoelectric bimorph. Jan Smits[7] is one of the more prolific authors in this field, and does an excellent analysis relating the conjugate variables of the system. His final result is very useful so we repeat it here:

$$
\begin{bmatrix}
\text{angle at tip} \\
\text{deflection} \\
\text{volume} \\
\text{piezo } charge
\end{bmatrix}
= A
\begin{bmatrix}
\dfrac{12L}{Kw} & \dfrac{6L^2}{Kw} & \dfrac{2L^3}{K} & \dfrac{6d_{31}BL}{K} \\[2mm]
\dfrac{6L^2}{Kw} & \dfrac{4L^3}{Kw} & \dfrac{3L^4}{2K} & \dfrac{3d_{31}BL^2}{K} \\[2mm]
\dfrac{2L^3}{K} & \dfrac{3L^4}{2K} & \dfrac{3L^5 w}{5K} & \dfrac{d_{31}BL^3 w}{K} \\[2mm]
\dfrac{6d_{31}BL}{K} & \dfrac{3d_{31}BL^2}{K} & \dfrac{d_{31}BL^3 w}{K} & \dfrac{Lw}{Ah_p}C
\end{bmatrix}
\begin{bmatrix}
\text{Moment at Tip} \\
\text{Force at Tip} \\
\text{Uniform Load} \\
\text{Piezo Voltage}
\end{bmatrix}
\tag{2.9.1}
$$

$$A = s_{11}^{Si} s_{11}^{P}(s_{11}^{P} h_{Si} + s_{11}^{Si} h_P) \tag{2.9.2}$$

$$B = \frac{h_{Si}(h_{Si} + h_P)}{(s_{11}^{P} h_{Si} + s_{11}^{Si} h_P)} \tag{2.9.3}$$

---

7. J. G. Smits and W. Choi, "The Constituent Equations of Piezoelectric Heterogeneous Bimorphs," IEEE Trans. Ultra. Ferr. Freq. Conf., **38**, 256 (1991).

$$C = \varepsilon_{33}^{T} - \frac{d_{31}^{2} h_{Si}(s_{11}^{P} h_{Si}^{3} + s_{11}^{Si} h_{P}^{3})}{K}$$

$$K = (s_{11}^{Si})^{2} h_{P}^{4} + 4 s_{11}^{Si} s_{11}^{P} h_{Si} h_{P}^{3} + 6 s_{11}^{Si} s_{11}^{P} h_{Si}^{2} h_{P}^{2} + 4 s_{11}^{Si} s_{11}^{P} h_{Si}^{3} h_{P} + (s_{11}^{P})^{2} h_{Si}^{4} \qquad (2.9.4)$$

where h is the thickness, s is the compliance (1/Young's modulus) of the silicon and the piezoelectric, L and w are the length and width of the cantilever, and ε is the permittivity at constant stress.

However, this and other[8] analyses do not include the electrodes and other films of the structure in the calculations, nor do they make provisions for the complexities of our standard geometry. Our analysis does include all the layers of the cantilever. Letting all the extra film thicknesses, and L_2 and L_3 go to zero, our results produce the same values as Smits' matrix. It is interesting to compare calculations which include all films of a structure and those which only include the bimorph.

Looking at the piezoelectric slab and otherwise using the standard geometry[9] equation (2.9.1) predicts a spring constant of 178 N/m. Including the electrodes we find k to be 270 N/m. It is reasonable to expect this large difference because spring constant is proportional to thickness cubed and the extra films add 1.2 μm to a 7 μm structure. The same is true for ZnO induced deflection. Without electrodes (2.9.1) predicts 53 Å/V, while with electrodes included, we calculate only 39 Å/V.

In addition, changes in position of the centroid affect MDD and actuator sensor coupling. To obtain a valid number for the centroid, again, all films must be included.

2.10 Summary

Other geometries can be analyzed using the approximations put forth in this chapter. The first step is to analyze the cantilever structure without including the actuator by applying the formulas given in Section 2.1. Second calculate MOI for the two regions, and find the spring constant modifier from equation (2.5.10).

8. M. R. Steel, F. Harrison, and P. G. Harper, "The piezoelectric bimorph: An experimental and theoretical study of its quasistatic response," J. Phys. D: Appl. Phys., **11**, 979 (1987).

9. L2 = L3 = 0, L1 = 180 μm, w = b / 2 = 85μm / 2, $t_{Si} = t_{ZnO}$ = 3.5 μm

We know that high doping in the ZnO region eliminates that region's contribution to the total $\Delta R/R$ and R. Therefore, $\Delta R/R$ for a regular cantilever can be scaled by the spring constant modifier to find the new $\Delta R/R$. MDD is inversely proportional to $\Delta R/R$ so it can be directly scaled. With the high doping in the ZnO region and reasonable centroid placement, it can be assumed that the coupling will be small (see section 2.8). Total induced deflection can be estimated by finding deflection and angle from Smits' matrix and geometrically extending those results to account for the compound geometry (section 2.6).

CHAPTER 3

Increasing the Speed of Imaging

3.1 Introduction

The electron beam in the Scanning Electron Microscope is scanned electronically at several meters per second over a large area whereas the tip in the Scanning Probe Microscope is scanned at several microns per second over a small area. These enormous differences limit the field of use for the SPM. It is imperative to correct this deficiency. The conventional piezotube scanner which exhibits slow mechanical response with a limited reach in the transverse (x-y) plane is the source of the problem. The scanning speed can be increased with an integrated actuator that acts only on the cantilever as described in this chapter. The scanned area can be increased with parallel arrays of cantilevers as described in Chapter 6.

High speed scanning has been previoulsy reported by Barrett[1], and Mamin[2]. These systems serve as powerful demonstrations but they aren't the complete solution. Barrett's system operated in a non-feedback mode and this required hard samples with little topography. Mamin's system scanned the sample with a low-displacement, high-resonance actuator and thus limited the mass and topography of the sample.

1. R. C. Barrett and C. F. Quate, "High-speed, large-scale imaging with the atomic force microscope," J. Vac. Sci. & Technol. B**9**, 302 (1991).

2. H. J. Mamin, H. Birk, P. Wimmer, and D. Rugar, "High-speed scanning tunneling microscopy: principles and applications," J. Appl. Phys. **75**, 161 (1994).

The scanning speed with constant force imaging is generally limited by the speed, or resonance, of the z-axis actuator. For high speed scanning the piezotube is replaced with a piezoelectric film deposited directly onto the cantilever. The expansion and contraction of this film when electric field is applied bends the cantilever along the z-axis to change the spacing between the tip and sample. For a sample with a given periodicity the scanning speed is directly proportional to the operating frequency of the z-axis actuator. The resonant frequency of the cantilever is a factor of 100 greater than that of the piezotube and this permits the scanning speed to be increased by a similar factor.

The scanning probes with film actuators are fast enough to provide zoom and pan imaging capabilities. The instruments have the feel on the SEM with improved spatial resolution. The increase in scanning speed, together with the increase in scanned area provided by the arrays, will facilitate and improve the performance in the fields of metrology[3], data storage[4] and lithography[5,6].

There is a penalty attached to the integration of actuators and sensors on the same cantilever beam. It is difficult to decouple the large signals required to drive the piezoelectric actuators from the small signals generated by the piezoresistive sensor. The optical lever is a fine solution for single cantilevers but it is not easily adapted to arrays. In a later chapter we introduce the interdigital cantilever as an interferometric optical sensor with excellent sensitivity that is easily adapted to our arrays.

Here we will discuss high speed scanning with a single cantilevers using three different sensors: 1) piezoresistive elements, 2) piezoelectric films, and 3) optical levers.

3. Y. Martin and H. K. Wickramasinghe, "Precision micrometrology with scanning probes," *Future Fab International*, **1**, 253-256 (1996).

4. H. J. Mamin, D. Rugar, "Thermomechanical writing with an atomic force microscope tip," Appl. Phys. Lett., vol. 61, 1003 (1992).

5. S.W. Park, H.T. Soh, C.F. Quate, and S.I. Park, "Nanometer scale lithography at high scanning speeds with the atomic force microscope using spin-on glass," Appl. Phys. Lett. vol. 67, 2415 (1995).

6. S. C. Minne, H. T. Soh, Ph. Flueckiger, and C. F. Quate, "Fabrication of 0.1 μm metal oxide semiconductor field-effect transistors with the atomic force microscope," Appl. Phys. Lett., vol. 6, 703 (1995.)

3.2 Parallel Cantilevers Operating Under Individual Feedback Control

In Chapter 1 we demonstrated that piezoresistive technology can be used to operate cantilever arrays imaging in the constant height, or open loop, mode. We scanned five cantilevers as a single unit with no provision for individual control of the tip-to-sample spacing. As an introduction to more complex systems we start by describing the operation of two cantilevers under simultaneous, individual feedback control. In most applications the motion of the individual cantilevers need only be controlled along the z-axis which is normal to the sample. In the x-y plane the array can be scanned as a single unit.

For the device shown in Figure 2.2.2 on page 26, our analysis predicts a DC deflection of 169 Å/V when we account for the multiple layers. The analysis of a uniform system where a single ZnO layer extends the full length of the beam has been carried out by Smits [7]. When his analysis is modified to account for the partial coverage of the beam with the layer of ZnO, it yields a deflection of 196 Å/V. The actual deflection of our device was 153 Å/V as measured by optical methods.

Non-uniformity and defects introduced by the fabrication process cause variation in the ZnO actuator response. However, these variations do not restrict imaging because the cantilever is operating in feedback where a constant force is always maintained. Calibration of the vertical scale in the image depends on the ZnO response, and varies from cantilever to cantilever.

In Figure 3.2.1 we display the frequency response for the cantilever shown in Figure 2.2.2. The total deflection of the cantilever is 1 micron for an applied field of $\pm 10^7$ V/m. For other geometries, we have obtained DC deflection sensitivity of 577Å/V or 4 microns for our typical ±35V operating range. In Figure 3.2.1 the tip deflection was measured optically.

In a conventional AFM either the cantilever or the sample is driven with a piezoceramic which bends to provide scanning in the lateral plane and extends to control the tip spacing in the z-dimension. When a laser is used to monitor the deflection of the ZnO cantilever system it is equivalent to the conventional AFM.

---

7. J. G. Smits, and W. Choi, "The Constituent Equations of Piezoelectric Heterogeneous Bimorphs," IEEE Transactions on Ultrasonics Ferroelectrics and Frequency Control, **38**, 256 (1991).

Increasing the Speed of Imaging **51**

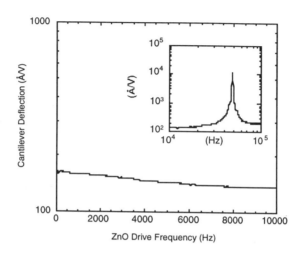

Figure 3.2.1   Cantilever tip deflection vs. ZnO drive frequency. The DC response is 150 Å/V. Maximum operating voltage +/- 35V.

On the other hand, when a piezoresistor is used to monitor the deflection the situation is more complex since the piezoresistor extends over the full length of the compound cantilever. The piezoresistor that lies beneath the ZnO layer generates an unwanted signal that is unrelated to the force on the tip. This signal can either be reduced by varying the doping profile in the piezoresistor during fabrication, or it can be compensated for during imaging.

When cantilevers are fabricated without heavy doping under the ZnO, it is straightforward to record the unwanted signal by stressing the ZnO film while the tip is freely suspended. We use this method to calibrate an amplifier in such a way that the unwanted signal is subtracted from the piezoresistor output during the imaging cycle.

This procedure for constant force imaging can be verified using a standard AFM where the motion of the tip is known accurately. The tip of the compound ZnO cantilever is placed in contact with the standard cantilever driven with a piezo tube. There is a one-to-one correspondence between the motion of the piezo-tube and the tip of the compound cantilever. One can use this method to verify that subtracting the response from the resistor under the ZnO yields a corrected signal from the piezoresistors that is a true measure of the tip deflection. This corrected signal is applied to the ZnO film as the feedback signal in the constant force mode of imaging.

We have used this method with our two-cantilever array to simultaneously acquire the multiple images shown in Figure 3.2.3. The sample is a one-dimensional grating with

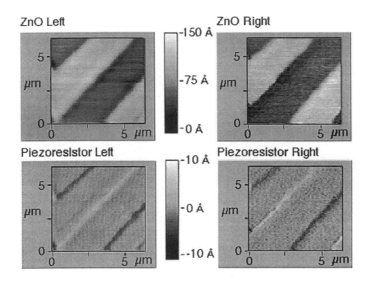

Figure 3.2.3   Parallel constant force images taken with two ZnO cantilevers operating simultaneously. The ZnO signal is the final image of the grating which has a 5 um period and 150 A step height. The piezoresistive output is the error signal. The flatness of the error signal, and positive and negative shifts at the grating step edges, indicate that constant force imaging is achieved.

a 5 μm period and a step height of 150Å. The ZnO signal records the topography (top images) while the piezoresistor output provides the error signal (bottom images). In the constant force mode the force between tip and sample should remain constant during the imaging cycle. The small error signal of Figure 3.2.3 (note the scale height) indicates that this condition has been satisfied.

The correction circuit provides a adequate method for constant force imaging. However, with our circuit, the corrected ZnO range is limited by non-linearities in the device. A better solution than correction of sensor-actuator coupling is to eliminate it through doping variation during fabrication.

Cantilevers with high doping beneath the ZnO eliminate sensor-actuator coupling and the need for a correction circuit. A schematic diagram of the side view of the cantilever showing the heavy implant is shown in Figure 3.2.4. The increased doping serves

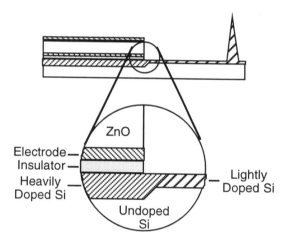

Figure 3.2.4   Schematic cross-sectional view of a piezoresistive cantilever with an integrated piezoelectric actuator. The piezoresistor lies on the surface of the silicon, separated from the ZnO actuator by silicon dioxide and silicon nitride. The piezoresistor doping extends to the apex of the tip.

two purposes.

First, increased doping under the ZnO region decreases the piezoresistive coefficient by 80%, which proportionally reduces the contribution to the sensor signal induced by the actuator. Second, as the doping increases, the resistance decreases. As a result, the absolute change in resistance from a given stress is reduced.

Reduced coupling permits imaging without the need for correction circuitry. An uncorrected parallel constant force image of an integrated circuit containing vertical topography of 2 μm is presented in Figure 3.2.6. The top images are formed from the

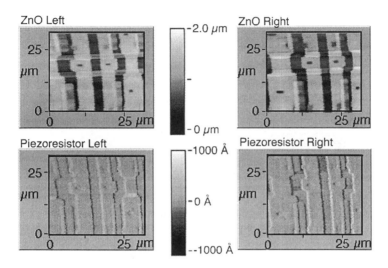

Figure 3.2.6   Parallel, constant-force AFM image of a microcircuit. The ZnO maintains constant force over two microns of topography. The error signals (piezoresistor) indicate that constant force is maintained.

voltages controlling the ZnO actuator and represent the surface topography, while the lower images are the piezoresistor signals which represent the error signals. For constant force imaging the error signals should be zero. The reduced scale on the error signal images shows that constant force is effectively maintained.

Hysteresis and linearity measurements of the cantilever are presented in Figure 3.2.7. These measurements were obtained by measuring the cantilever deflection with an optical lever system. The optical system is calibrated to the out of contact movement of the cantilever in the following manner. First the tip is brought close to the surface of a calibrated piezotube. Next the voltage to the ZnO is increased until the tip strikes the sample. The piezotube is then retracted from the tip by a known amount. The amount of additional voltage to the ZnO required to bring the tip back into contact with the surface provides a calibration for tip movement in air versus change in ZnO voltage. The cantilever is then deflected by the ZnO in air while the output of the laser

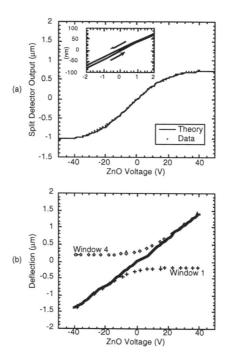

Figure 3.2.7   *a) Hysteresis curve for the ZnO actuator. The points represent the data while the solid line is a simulation of a Gaussian beam passing over a split photodiode. Inset: Expanded view of the data around zero applied voltage showing 20 nm maximum hysteresis. b) Linearity plot for the ZnO actuator. The solid line is a composite curve of four photodiode positions representing the overall linearity curve for the actuator. The points of the first and last windows used to construct the overall plot are included. This method was required because the cantilever displaced the beam beyond the linear range of the photodiode.*

is monitored. The slope of the response combined with the previous result gives the change in laser output for a given displacement of a freely suspended tip.

Figure 3.2.7a shows the output of a split photodetector when the ZnO voltage is swept from -40V to 40V and then back to -40V. The inset of figure Figure 3.2.7a shows an expanded view of the hysteresis loop around zero volts applied voltage. The maximum hysteresis of the ZnO in this voltage range is 20 nm. Unfortunately, it is not possible to determine the linearity of the actuator from Figure 3.2.7a. The split photodiode detector used to measure the position of the beam reflected off the cantilever has a limited range for which the output is linear with respect to cantilever deflection. This limitation occurs because the size of the reflected beam is finite and must cover an adequate portion of both sides of the detector. In our microscope, the extent of the linear range covers roughly 1 μm of deflection and is insufficient to characterize the entire range of the ZnO actuator. The sigmoidal shape of Figure 3.2.7a reflects the gaussian beam profile of our laser diode. We have also simulated a gaussian beam traversing a split photodiode and plotted the results with the data in Figure 3.2.7a.

In order to measure the linearity of the actuator, we manually translated the linear window of the detector in discrete steps so that it covered the entire range of the reflected laser beam. This provided us with a series of curves, each of which has only a small linear range. We then extracted and shifted the linear data from each curve to construct the overall linearity curve. The data was shifted such that overlapping voltages in adjacent windows had the same deflection.

Two such windows, which represent the initial and final data curves used for constructing the overall plot, along with the final construction, are plotted in Figure Figure 3.2.7b. In these windows, the detector position was adjusted so that response is linear at the maximum excursion of the cantilever. Once the beam is completely deflected to one side of the photodiode, the detector cannot provide information about the cantilever deflection and the curves flatten. The composite curve of figure Figure 3.2.7b shows excellent linearity of the device over its entire range of 3 μm.

3.3 High Speed Imaging using the Piezoresistive Sensor

The piezoresistive cantilever is usually measured at DC with a Wheatstone bridge. However, in our generation of devices, the DC bridge does not measure true cantilever deflection because the topology of our devices permits electrical cross-talk between the ZnO signal and piezoresistor signal. This capacitive coupling between the piezoresistive sensor and the ZnO actuator is shown in Figure 3.3.1. The y-axis of Figure 3.3.1 is a normalized measure of the apparent topography generated in the piezoresistive signal from an AC voltage applied to the ZnO. Ideally this curve should have a constant value of 0 μm/V (except at the cantilever resonance). We have found that the problem of electrical coupling can be solved by measuring the response

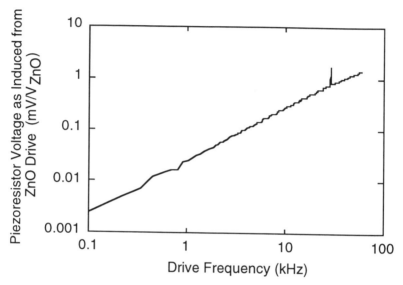

Figure 3.3.1 *Piezoresistor response versus ZnO drive frequency for a cantilever*
not in contact with a surface. Since the actuator and sensor are
integrated on the same cantilever, the piezoresistive signal is
influenced by capacitive coupling from the ZnO drive signal. The
piezoresistor remains unstressed since the end of the cantilever is
free. A 0.3 mV signal from the piezoresistor corresponds to a
cantilever deflection of about 1 um.

of the piezoresistor with a lock-in amplifier. By modulating the piezoresistor at a frequency well above the imaging bandwidth, the response is immune to the effect of electrical coupling and sensitive to deflection.

A schematic of an AFM containing an integrated actuator and sensor for high speed imaging is shown in Figure 3.3.2. The piezoresistor is placed in series with a resistor, R, to form a voltage divider. The other end of each resistor is then driven with an AC signal, $V\sin(\omega t+\phi_1)$ and $-V\sin(\omega t+\phi_2)$. The divider is balanced by alternately adjusting two parameters. First, R is matched to the resistance of the piezoresistor. Second, the relative phase between the driving signals (ϕ_1 and ϕ_2) is set so that it differs by roughly 180°. The output of the divider with frequency component f_o ($f_o=\omega/2\pi$) is then minimized to the microvolt level by further adjusting R and either ϕ_1 or ϕ_2. It is

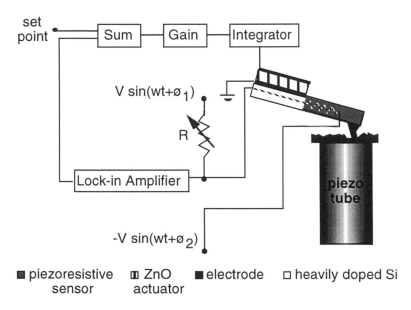

set point

Sum — Gain — Integrator

$V \sin(wt+\phi_1)$

R

Lock-in Amplifier

piezo tube

$-V \sin(wt+\phi_2)$

■ piezoresistive   ▥ ZnO   ■ electrode   □ heavily doped Si
    sensor   actuator

*Figure 3.3.2   Schematic of electronic correction circuit. The lockin amplifier
                measures the piezoresistance in the voltage divider.*

not sufficient to create an exact 180° phase difference by simply inverting one of the signals since parasitic capacitance and inductance add unknown phase shifts. In this work, V is typically a few volts, R ranges from 3-4 kΩ, and f_o is set to a frequency of 130 kHz. When operating in constant force mode, the signal generated by the cantilever deflection is detected with the lock-in amplifier and the output of the feedback circuit is connected to the upper electrode of the ZnO.

The maximum imaging bandwidth of the cantilever and lock-in detection system is determined with the following procedure. The cantilever is placed in contact with a fixed surface and the ZnO actuator is driven with a sine wave at various frequencies. The amplitude and phase response of the piezoresistor are recorded with an additional lock-in amplifier. Figure 3.3.3 shows the amplitude and phase response of the cantilever described in the previous section. If we allow for a phase margin of 45°, the self-actuated cantilever and lock-in piezoresistor detection system provide roughly 6 kHz of imaging bandwidth. The resonance of the cantilever when in contact with the sample surface is near 44 kHz. Due to limitations of the electronics used to drive the piezoresistor divider, we were not able to balance the divider for frequencies above

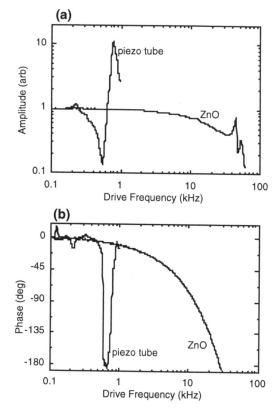

Figure 3.3.3 (a) Amplitude and (b) phase response of the piezoresistor as a function of ZnO drive frequency. With a 45 degree phase margin, the maximum imaging bandwidth is 6 kHz. For comparison, the response of a cantilever driven with a 2-inch piezo tube.

130 kHz. The gradual roll off of the phase below the cantilever resonance indicates that the imaging bandwidth is limited by the detection electronics rather than the mechanical response of the cantilever. An imaging bandwidth equal to the cantilever resonant frequency should be achievable if the divider is driven at a frequency much larger than the resonance.

Also shown in Figure 3.3.3 is the piezoresistor response for the case where the sample is driven with a piezotube. The length of the tube is 2 inches, a commonly used size for applications requiring scans up to 100 μm. The sharp 180° phase shift near 600 Hz is associated with a lower resonance of the piezo tube.

An image of an integrated circuit taken with a tip velocity of 3 mm/s in the constant force mode is shown in Figure 3.3.4. A line profile along the dotted line is shown

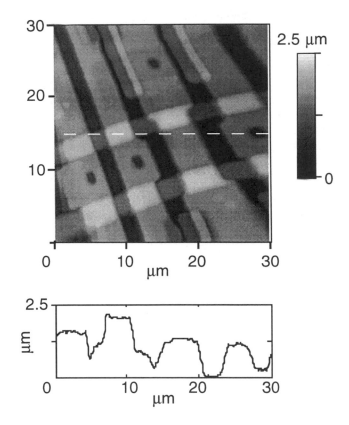

Figure 3.3.4 512 by 512 pixel image acquired at 3 mm per second tip speed. The sample is an integrated circuit with metal lines and contact holes. Also shown is a line profile taken at the position of the dotted line in the image.

below. The integrated circuit consists of metal lines and contact holes and contains vertical steps 2 μm in height. The ZnO drive signal required to bend the cantilever over the sample topography was less than ±30 V. The sample was imaged by raster scanning over a 30 μm x 30 μm area using a 1 inch long piezo tube with a fast scan rate of 50 Hz. The high resolution image of 512x512 pixels was acquired in roughly 15 seconds. To circumvent resonances in the tube, the fast scan direction was driven with a sine wave while the slow scan was ramped with a triangle wave. A video acquisition system was used for the fast data acquisition. Given an x and y input, this system digitizes the z output and converts it to a video signal so that the image can be displayed on a monitor in real-time. The images are then captured on video tape and downloaded to a computer for analysis. By adding an interlace to the slow scan and changing the pixel resolution to 128x128, it was possible to acquire images at a rate of several frames per second. The fast visual feedback with this scanning rate makes it easy to adjust parameters such as position, zoom, and rotation, giving the AFM the feel of a scanning electron microscope (SEM).

A detailed surface plot taken from the image in Figure 3.3.4 is shown is Figure 3.3.5.

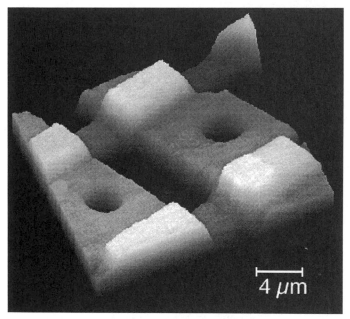

Figure 3.3.5   Surface plot of the image in Figure 3.3.4

Fine surface morphology revealed in this plot and in the line plot of Figure 3.3.5 shows lateral features of roughly 100 nm.   The lateral resolution is commensurate with collecting 512 pixels over a 30 µm line scan and does not change if the same image is taken ten times more slowly. The vertical resolution of this cantilever in a 10 Hz to 7 kHz bandwidth is 60 Å. With a shorter cantilever (180 µm ZnO base plus 240 µm piezoresistor), the resolution is increased to 35 Å.

3.4 Imaging using the ZnO as the Sensor

In this section we describe a method for constant force imaging in which the ZnO cantilever is used as both an actuator and a sensor. For this mode the same cantilever and experimental configuration can be used for both static and dynamic applications.

By employing the ZnO as both an actuator and a sensor, the design of the cantilever can be significantly simplified by eliminating the need for the piezoresistor.

The cantilever consists of a rigid actuator supporting a flexible sub-cantilever which interacts with the surface (see Figure 3.4.1). The base of the cantilever is the ZnO actuator and this region is roughly 10 times stiffer than the cantilever extension. The extension that is attached to the cantilever can be made of any material. In our design it is silicon with a built in piezoresistor. This resistor can be used as a piezoresistor for an alternative method of sensing deflection, but since we now use the ZnO as both the sensor and actuator, the silicon resistor is freed to be used as a bias path for lithography, heating, or tunneling in future experiments. This geometry provides the versatility of a robust actuator with a soft cantilever for probing the surface of the sample.

Our method for constant force imaging is based on the second resonance of the cantilever. The second resonance is a flexural mode with a node at the tip, and does not appear when the tip of the cantilever is free. When the tip is placed in contact with a surface, the second resonance appears and its amplitude varies as a function of force. The vibrating tip in contact with the surface transmits energy into the sample to a degree that depends on the average force that the tip exerts on the sample. The power dissipated by the tip is easily measured by monitoring the current into the ZnO film. Since the drive voltage is constant, the current is proportional to the ZnO admittance. The admittance will vary as the quality of the resonance changes in order to conserve power between the electrical and mechanical systems. This admittance variation is used for the deflection signal in the feedback loop.

Figure 3.4.1 is a schematic of our experimental setup. When using the ZnO as both the sensor and actuator the relevant electronics are shown in the dashed box labeled "ZnO Admittance". The dashed box labeled "Piezoresistor" are the electronics for using the piezoresistor as the sensor. A comparison of the piezoresistive signals and the ZnO signals is presented later.

Figure 3.4.2b is a plot of piezoresistor amplitude versus ZnO drive frequency for various tip sample spacings. The line in the lowest part of the plot corresponds to the tip being far away from the sample. As the lines progress upward, the tip is moved towards the sample by 150Å per line. The resonant peak at 132kHz represents the third mode of the cantilever. As the tip moves towards the sample the amplitude is reduced because the tip intermittently strikes the sample. This can be seen in the middle region of Figure 3.4.2b. The resonance at 132kHz vanishes completely when the tip comes into contact with the surface for the entire cycle. At this point, the boundary condition at the end of the cantilever changes from a free unrestricted movement to

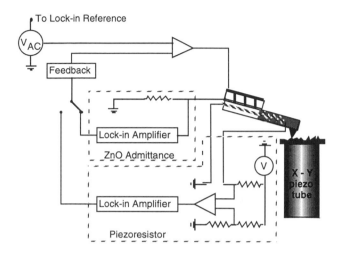

Figure 3.4.1   Schematic of experimental setup. The electronics relevant to using the ZnO as both the sensor and the actuator are in the dashed box labeled "ZnO Admittance". The dashed box labeled "Piezoresistor" contains the circuit for the piezoresistor as the sensor.

constrained motion due to the Hertzian contact with the surface. This change creates a second resonance at 102kHz. Increasing force after tip/sample contact changes the amplitude, the Q of the resonance, and the resonant frequency.

Figure 3.4.2a is the same plot as Figure 3.4.2b except the admittance of the ZnO is measured rather than the stress in the piezoresistor. The coupling between the mechanical system and the ZnO is greater at higher order modes due to the increased curvature in the ZnO. This allows us to monitor changes in the ZnO admittance at higher order modes, which was not possible at the fundamental. In Figure 3.4.2a the variation at 120kHz is an external electromagnetic interference signal that is picked up by interconnecting wires, and is unrelated to the cantilever.

Figure 3.4.2c shows the fundamental, or first, resonance of the cantilever measured with the piezoresistor at 35.5kHz. We note that the fundamental resonance can be used for intermittent contact imaging with the piezoresistor as a sensor and the ZnO as an actuator.

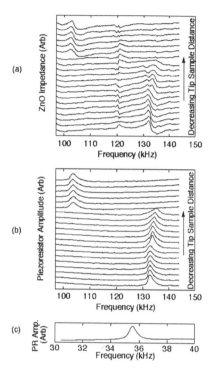

*Figure 3.4.2   Plot of the second and third cantilever resonance frequencies using
the (a) zinc oxide and (b) piezoresistor. Each line represents a 150 Å
movement of the tip toward the sample. Curves at the bottom of each
plot are out-of-contact while those at the top are in-contact. (c) Plot
of fundamental mode using piezoresistor sensor.*

To examine the degree of change in the resonant peaks with force we excite the ZnO
at a given resonance, use a lock in amplifier to measure the piezoresistor and the ZnO
admittance, and vary the tip-to-sample spacing. The results are presented in
Figure 3.4.3 as standard force curves. In Figure 3.4.3 negative tip sample distances
represent the tip out of contact with the sample. The sharp dip in the force curves corresponds to sticking due to the meniscus between the tip and sample. The signals presented in Figure 3.4.3 may be combined with noise data (not shown) to obtain
minimum detectable deflection.

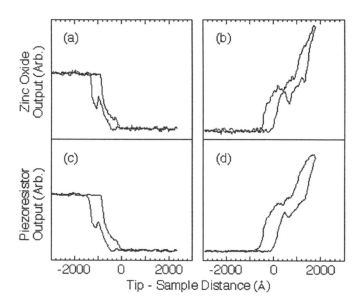

Figure 3.4.3   Force curves. The left-hand column represents intermittent-mode measurements while the right shows in-contact data. (a) Intermittent contact; ZnO sensor. (b) In-contact; ZnO sensor. (c) Intermittent contact; piezoresistive sensor. (d) In-contact; piezoresistive sensor. Negative distances represent the tip out-of-contact with the sample.

Figure 3.4.3a was performed with an excitation of 132 kHz (3rd order mode) and a tip amplitude of 1000Å (0.1 V to ZnO). The admittance is detected by measuring a current through a 10 kΩ resistor in series with the ZnO, and corresponds to the cantilever deflection. The form of this curve suggests an intermittent contact mode. There is no change in admittance while the tip is far from the sample. As the tip approaches, it begins to strike the surface, and the amplitude of the resonance decreases. Once the tip is in complete contact with the surface, the admittance no longer changes.

In Figure 3.4.3b the cantilever was excited at the second (102kHz) resonance, with the ZnO drive again at 0.1 V, and the ZnO admittance was monitored. The form of this curve suggests a contact mode. When the tip is away from the sample the response is zero. Once the tip contacts the surface the response increases due to the increased amplitude of the resonance. Since there is no resonance present at 102kHz when the tip is away from the sample, an excitation of 0.1 V to the ZnO causes only a 30Å deflection at the tip. The range of the force curve shown extends over 2000Å, which is considerably greater than the on, or off, resonance amplitude of the tip. This indicates that this is truly a contact mode, not a modified intermittent contact mode.

Figure 3.4.3c & d correspond to measurements taken in the same manner as Figure 3.4.3a & b except the piezoresistor is used as the sensor instead of the ZnO. From the traces in Figure 3.4.3, it is apparent that the noise in the piezoresistor (PR) traces (c, d) is less than the noise in the ZnO traces (a, b) for both intermittent contact and contact modes. The minimum detectable deflection, where the cantilever response corresponds to the RMS noise, for the four modes in a 100 Hz bandwidth are: a) intermittent contact/ZnO = 30Å, b) contact/ZnO = 40Å, c) intermittent contact/PR =20Å, d) contact/PR = 30Å. The minimum detectable deflection for intermittent contact using the fundamental mode and DC constant force is 10Å in a 100 Hz bandwidth.

In the initial design of this cantilever we were planning to use the ZnO only as an actuator, we did not anticipate its use as a sensor. In order to facilitate the process of fabrication, we designed the ZnO film in such a way that the major portion covered the die with a minor portion covering the cantilever. The major portion, of course, is insensitive to cantilever vibration. The area of the film on the die is 26 times larger than the area of the film on the cantilever. It is feasible in a new design to eliminate most of the film on the die and this should improve the sensitivity by an order of magnitude.

Images are readily obtained from these four new probing techniques, and are presented in Figure 3.4.4. The images in Figure 3.4.4 are in the same order as that of Figure 3.4.3: a) intermittent contact/ZnO, b) contact/ZnO, c) intermittent contact/PR, d) contact/PR. The sample is a two dimensional gold grating with a period of 1μm and height of 1000 Å. All of the images were taken with feedback to the ZnO actuator.

3.5 High Speed Imaging with the Optical Lever Sensor

In a previous section, we found that the scan speed could be increased an order of magnitude by integrating a thin layer of ZnO on the base of a piezoresistive cantilever. The cantilever was bent to follow sample topography by applying a voltage across the ZnO while the sample force was detected by measuring the piezoresistor. In that study, both the imaging bandwidth (6 kHz) and the resolution (~60 Å) were limited by

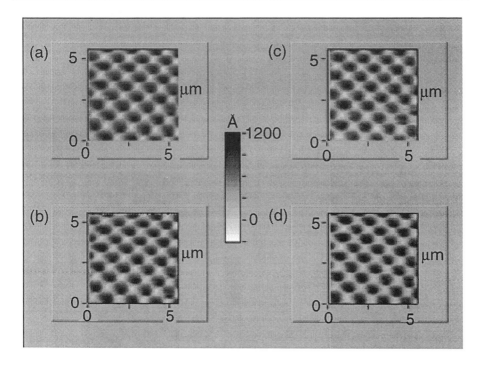

Figure 3.4.4   Images taken with feedback to the ZnO sensor. (a) Intermittent contact; ZnO sensor. (b) Contact; ZnO sensor. (c) Intermittent contact; piezoresistive sensor. (d) Contact; piezoresistive sensor.

complications in measuring the piezoresistor. We expect that these results can be improved by optimizing the cantilever design and refining the detection electronics. However, the susceptibility to unwanted electrical interference of a detector consisting of an electrical loop can make it difficult to obtain large bandwidths with high resolution.

Here we present the extension of high speed imaging with an integrated actuator to include an optical lever sensor. The optical lever is capable of sub-angstrom vertical resolution and is not influenced by electrical signals that control the integrated actuator. In a standard system, constant tip-to-sample force is achieved by translating either the cantilever or the sample vertically with a servo loop that maintains a constant cantilever deflection. The optical lever is used to measure the deflection angle of the cantilever. In ordinary circumstances, this angle only changes when a force is acting on the tip. Thus the optical lever gives a true measure of the force. In a system where the actuator is mounted on the cantilever, however, motion in the vertical direction is achieved by changing the angle of the cantilever. This is a problem that must be addressed when the optical lever is used for detection. This situation is depicted in Figure 3.5.1 for three cases where: (a) the tip-to-sample force is near zero, (b) the can-

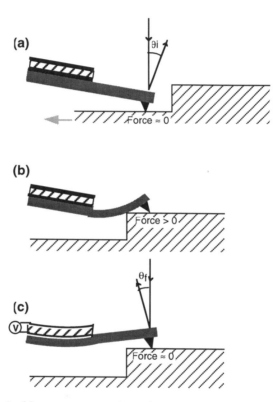

Figure 3.5.1   Optical lever compensation schematic.

tilever is strained by a topographical step, and (c) a voltage is applied to the ZnO to relieve the strain. The angle of the reflected beam in (a) is different from the angle in (c) although the force is the same. As we will discuss, the signal measured by the optical lever is a function of the ZnO bending in the absence of a force on the tip. It is a simple matter to construct a circuit that subtracts this unwanted signal from the servo loop. We have found that a first order linear correction applied to a 720 µm long cantilever yields a vertical range where constant force is possible. Using this correction procedure, we present a 100 μm x 100 μm image scanned with a tip velocity of 10 mm/s, as well as a high resolution image of the granular structure of gold scanned at 0.5 mm/s. In the case of the high resolution image, the tip speed was limited by the scanning device.

A schematic of our atomic force microscope (AFM) is shown in Figure 3.5.2. The silicon cantilever can be displaced vertically up to 4 μm using a layer of ZnO located at the cantilever base. Because the spring constant is proportional to the thickness cubed, and the base is twice as thick as the remainder of the cantilever, most of the bending will occur in the thinner portion when the tip is deflected. Therefore we can effectively uncouple the deflection of the cantilever caused by applying a voltage across the ZnO from the deflection caused by physically displacing the tip with a sample. As a result, if the tip is in contact with a surface while the ZnO voltage is modulated, the deflection signal from the photodiode will consist of two angular components: the strain-induced angle (depicted in Figure 3.5.1b), and the ZnO-induced angle (Figure 3.5.1c). To obtain constant force, the servo loop must detect only the strain-induced angle.

To eliminate the ZnO-induced angle, we sum the output of the photodiode with the signal that controls the ZnO (see Figure 3.5.2). In order to calibrate the correction circuit, the ZnO is modulated while the tip end of the cantilever is free and the gain of the photodiode output is adjusted such that the corrected deflection is nulled. At this point, the servo loop cannot distinguish between the scenario depicted in Figure 3.5.1a and c. Since the cantilevers used in this experiment contain an integrated piezoresistor, constant force can be verified by modulating the sample in the vertical direction with the piezo tube while the ZnO actuator and corrected photodiode signal are in the feedback loop.

The piezoresistance represents strain in the cantilever and is a direct measure of the force at the tip. Our cantilever has been fabricated such that the geometry and doping profiles cause the piezoresistor to respond to forces on the tip, but not to the strain induced by ZnO movement.[8] This configuration provides an accurate measure of the tip-to-sample force regardless of the degree of ZnO actuation.

---

8. S. C. Minne, S. R. Manalis, A. Atalar, and C. F. Quate, "Independent Parallel Lithography using the Atomic Force Microscope," Presented at the Fourth International Conference on Nanometer Scale Science and Technology (NANO 4), Beijing, China, (September 1996). J. Vac. Sci. Technol. B, vol. 14, no. 4, p. 2456 (1997)

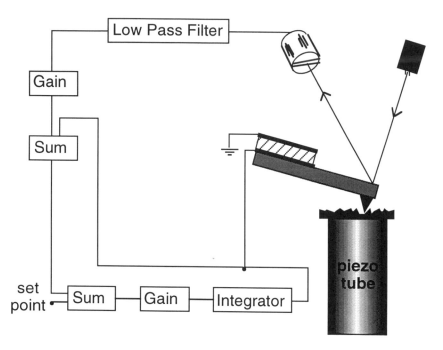

Figure 3.5.2   Schematic of high speed optical lever setup

Figure 3.5.3 shows the output of the piezoresistor as a function of the relative piezo tube position when the photodiode signal is corrected (solid) and uncorrected (dashed). When the correction procedure is implemented, the piezoresistive signal is constant over a piezo tube range of 1 μm, indicating that constant force is maintained. When the correction scheme is omitted from the servo loop, the piezoresistance varies with the tube position, indicating that the tip/sample force is not constant. For piezo tube positions greater than 1 μm, the laser beam was deflected out of the linear range of our photodiode.

The frequency response of the actuator, sensor, and servo loop is determined by modulating the setpoint while the tip is in contact with a surface (see Figure 3.5.2). First, the integral gain and time constant of the servo loop are increased to just below the point where the system becomes unstable. The modulation frequency of the setpoint is then varied while the output of the integrator (ZnO signal) and input to the gain/integrator (error signal) are recorded. The amplitude and phase response of the ZnO are shown in Figure 3.5.4a, and the error signal is shown in Figure 3.5.4b for a 570

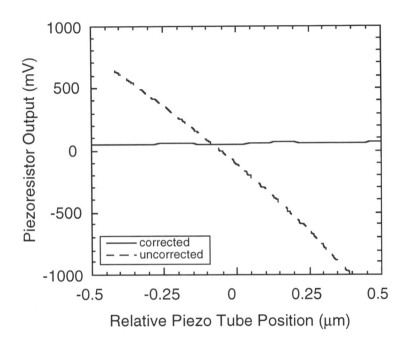

*Figure 3.5.3   Graph of piezoresistor output versus vertical position of the piezo
tube when the photodiode signal is corrected and uncorrected. The
piezoresistive signal obtained with correction indicates that constant
force is maintained over a one micron vertical range of the piezo tube
position. The piezoresistor is implanted in the region of the cantilever
not covered by ZnO and gives a direct measure of the force applied at
the tip.*

μm long cantilever. By adding a 12 dB low pass filter at 100 kHz, we increased the
integral gain while still maintaining stability of the servo loop. The resulting bandwidth is 33 kHz for a 45° phase shift of the error signal. The mechanical resonance at
77 kHz (shown in Figure 3.5.4) causes the system to oscillate if the integral gain is
increased further.

Large scale imaging in the constant force mode with a tip velocity of 10 mm/s is demonstrated in Figure 3.5.5. The sample was constructed by patterning circles on a 1000
Å thick gold film deposited on a Nb doped $SrTiO_3$ substrate. Images were obtained

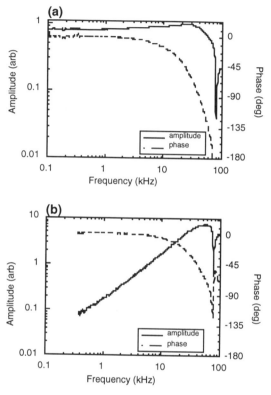

Figure 3.5.4   Frequency response of the servo loop shown in Figure 3.5.2. The input to the ZnO actuator is shown in (a) and the input to the gain-integrator stage (error signal) is shown in (b). The system is driven by modulating the setpoint with a variable frequency sine wave. With a 45 degree phase margin, the imaging bandwidth is 33 kHz. Note the mechanical resonance at 77 kHz.

by raster scanning the sample over an area of approximately 100 μm x 100 μm using a 2 inch long piezo tube with a fast scan rate of 50 Hz. A complete image consisting of 512 scan lines was acquired in under 15 seconds. In order to verify that constant force mode was being used, the cantilever stress was monitored with the piezoresistor at tip speeds up to ~3 mm/s. Electrical interference between the piezoresistor and the ZnO actuator was eliminated by measuring the piezoresistor with a lock-in technique. Limitations in our electronics prevented the use of this technique at higher speeds. To

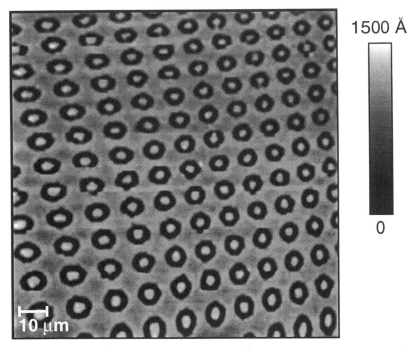

1500 Å

0

Figure 3.5.5   Constant force image acquired at 10 mm per second scan speed. The
sample is patterned gold on Nb doped SrTiO $_3$. The 512 line image
covers an area of approximately 100 by 100 um and was produced in
less than 15 seconds. The image is slightly distorted by the nonlinear
response of the piezo tube scanner.

circumvent resonances in the tube, the fast scan direction was driven with a sine wave
while the slow scan was ramped with a triangle wave. A video acquisition system was
used for the fast data acquisition.

Using the same scanning system, we reduced the scan size to the micron scale and
increased the scan rate to 200 Hz. Since many frames per second could be acquired at
this rate, we were able to center and zoom-in on a single feature with real time visual
feedback. Figure 3.5.6 shows a section of a 1.25 μm x 1.25 μm image revealing the
granular structure of the gold taken at a tip velocity of 0.5 mm/s. Although our servo
loop is capable of faster tip velocities, a mechanical resonance of our piezo tube just
above 200 Hz limited the scan rate. Since tip speed is a function of both scan rate and

Figure 3.5.6   Granular structure of gold taken with a tip speed of 0.5 mm per second. The scan area is 0.5 by 0.5 um and the vertical resolution is 40 Å.

scan size, the tip speed was limited to roughly 0.5 mm/s for scan sizes on the order of 1 μm². In a 10 kHz bandwidth, the measured electrical noise of the system corresponds to a deflection of a few angstroms.

3.6 Dynamic Imaging Modes

A constant force between the tip and sample can be maintained by incorporating the integrated actuator and sensor in a feedback loop. The feedback circuitry and operation is identical to systems where a piezo tube is used in place of the ZnO actuator. A key advantage of systems containing integrated sensors and actuators is that multiple cantilevers can be operated in parallel while maintaining a constant force. Demonstrations have shown that a modular cantilever design can be replicated to produce an array of 50 cantilevers with a 200 um pitch. Constant force with multiple cantilevers within the array is achieved with a computer controlled array of analog feedback channels. Parallel operation increases the overall imaging throughput in proportion to the number of cantilevers in the array. In addition, arrays can be used to directly image large surface areas while maintaining the high lateral resolution associated with scanning probe microscopes. This is of particular importance for the inspection of semiconductor surfaces where the dimensions of surface contamination or defects can be below the wavelength of visible light thereby making optical inspection difficult.

Images can also be obtained in a dynamic mode where the cantilever is mechanically driven at resonance such that the tip lightly strikes the sample surface at the end of each oscillation. This mode of imaging is often referred to as tapping, or intermittent contact, and is one of the most common techniques used in industry today.[9] Under ambient conditions, a sample surface contains a fluid layer that is a few nanometers thick. Tapping mode is advantageous for most imaging applications because problems associated with friction, adhesion, and electrostatic forces are alleviated. The tip is alternately placed in contact with the sample and then lifted off the sample to avoid dragging the tip across the surface. The non-destructive nature of the tapping mode is particularly useful for imaging soft biological samples. For applications in the semiconductor industry, the low lateral force between the vibrating tip and sample allows the roughness of silicon to be measured with angstrom resolution.

In tapping mode, sample topography is determined by measuring the oscillation amplitude of the cantilever. For example, as the vibrating cantilever is scanned over an upward step, the amplitude is reduced in proportion to the step height. Likewise, the oscillation amplitude will increase as the cantilever is scanned over a downward step. The cantilever is typically resonated at an amplitude of ~100 nm by oscillating the cantilever die with a piezoelectric slab. In order to track topography larger than the oscillation amplitude, a vertical actuator is used to translate the cantilever, or sample, such that the amplitude remains constant.

The ZnO actuator can be used to both drive the entire cantilever at resonance and offset the cantilever deflection such that the oscillation amplitude remains constant. The amplitude of the flexible portion of the cantilever is measured with the piezoresistive sensor by using a Wheatsone bridge and an RMS-to-DC converter. A diagram of this system is shown in Figure 3.6.1. The cantilever offset deflection is obtained by connecting the output of the integrator to the bottom ZnO electrode while the drive waveform is applied to the top electrode. The drive waveform consists of a sine wave with a frequency of a few tens of kilohertz (depending on cantilever dimensions) and an

9. Tapping Mode Imaging: Applications and Technology, Digital Instruments, Santa Barbara, CA.

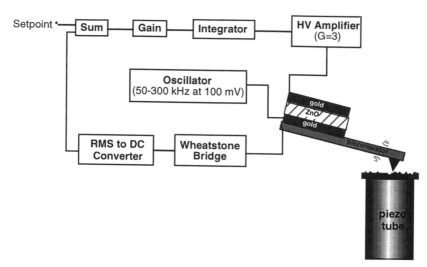

Figure 3.6.1 *Schematic for tapping mode operation with a cantilever incorporating a ZnO actuator and a piezoresistive sensor. The bottom ZnO electrode is used to drive the cantilever at resonance while the top electrode is used to bend the cantilever over sample topography such that the oscillation amplitude remains constant.*

amplitude of a few hundred millivolts. A high voltage amplifier is placed between the integrator and ZnO in order to produce a bipolar voltage swing of 35 V which corresponds to a few microns of vertical tip displacement. The output of the integrator represents the sample topography and the output of the RMS-to-DC converter is the feedback error signal.

A plot of the cantilever RMS amplitude displacement versus drive frequency for three different tip/sample separations is shown in Figure 3.6.2. The first curve is measured when the cantilever is far from the sample surface. When the separation is adjusted such that the tip comes within a few hundred angstroms of the surface, Van der Waals forces between the tip and surface create an attractive force gradient. The presence of this force gradient modifies the effective cantilever spring constant and the resonant frequency is shifted. This effect is shown by the second curve of Figure 3.6.2. At this tip-to-sample separation, the surface of samples can be imaged by tracking the frequency, amplitude, or phase shift of the cantilever. This technique is often referred to as the non-contact mode of imaging and in many cases provides lower resolution and less stability than other imaging modes because the Van der Waals forces are

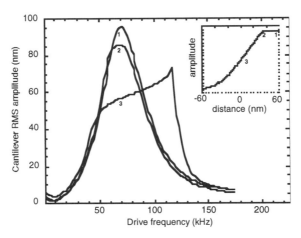

*Figure 3.6.2   RMS amplitude of cantilever displacement versus drive frequency.
Curve 1 is measured when the tip is far from the sample surface;
curve 2 is for a tip a few hundred Angstroms from the surface in the
non-contact regime; curve 3 is for a tip tapping the surface. The RMS
amplitude versus tip-to-sample is plotted in the inset.*

located near the sample surface and are substantially weaker than forces used by contact or tapping modes. If the tip-to-sample separation is further reduced, the tip will begin to strike the surface at the end of each oscillation and the amplitude will be reduced by an amount proportional to the offset of the tip towards the surface. This is the regime where tapping mode images are acquired. It is often useful to plot the cantilever amplitude at a fixed drive frequency as a function of the tip-to-sample separation (see inset of Figure 3.6.2). For non-contact imaging, the drive frequency should be set slightly below resonance such that when the tip encounters an attractive force gradient, the amplitude will increase as the tip approaches the surface. The setpoint of the feedback should be adjusted such the amplitude is fixed at the steepest portion of this nonlinear region. For tapping mode, the cantilever should be driven at resonance, and the amplitude should be kept as large as possible within the linear region. This minimizes the energy transfer between the oscillating cantilever and sample surface. An image of a two dimensional grating taken in the tapping mode is shown in Figure 3.6.3. Topography (integrator output) is plotted in Figure 3.6.3a and the error signal (RMS-to-DC output) is plotted in Figure 3.6.3b. Note that the vertical scale of the error signal is much smaller than the vertical topography scale. This indicates that

Increasing the Speed of Imaging   **79**

(a)   (b)

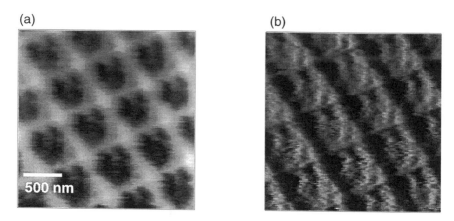

Figure 3.6.3 Tapping mode image of a gold diffraction grating. (a) Integrator output representing topography with a vertical scale of 100nm. (b) Cantilever displacement (error signal); vertical scale 10 nm.

the feedback is successfully maintaining a constant amplitude. Tapping mode has proven to be the most desirable imaging mode for many samples.

CHAPTER 4

Cantilevers with Interdigital Deflection Sensors

4.1 Introduction

We will present in this chapter an optical method for measuring the cantilever deflection that is suitable for arrays. It consists of a integrated, deformable, diffraction grating etched into the cantilever beam. In the new sensor several sets of rectangular fingers mesh together to form a deformable diffraction grating. This type of interference sensor is commonly known as the 'interdigital' cantilever. The intensity of the light diffracted into the higher orders by this grating gives us a measure of the deflection of the set of fingers attached to the beam that carries the tip. This is in contrast to the optical lever where the position of the optical beam is used to measure the deflection.

The new technique offers two primary advantages over the optical lever: First, the sensitivity is improved by an order of magnitude because intensity measurements are immune to pointing noise of the laser and mechanical vibrations of the photodetector. Second, alignment requirements of the photodetectors are reduced. These advantages make interferometric detection a viable technique for operation in cantilever arrays.

Operationally, the technique requires only an illumination source and a standard photodiode, yet it achieves a resolution that is comparable to state-of-the-art interferometric sensors. Simplicity allows the interdigital cantilever to be used in most optical lever AFMs without modification.

Micromachined diffraction gratings are used in many micro-optical systems including high resolution displays.[1] The idea of integrating a diffraction grating onto the cantilever to determine its deflection was suggested by Prof. Abdullah Atalar of Bilkent University while searching for a high resolution deflection sensor for arrays of cantilevers.[2] The interdigital cantilever alleviates the task of critically aligning individual photodiodes to an array since intensity rather than position of a diffracted beam is measured.

4.2 Theory of Operation

In order to understand the operation of the interdigital detection scheme, it is helpful to review optical diffraction from a one-dimensional grating. When a reflective grating is illuminated with coherent, monochromatic light, the majority of the light is reflected back towards the source (see Figure 4.2.1) This is called the 0th order diffracted mode. If alternating fingers are vertically displaced by $\lambda/4$ where λ is the illumination wavelength, the 0th order mode is cancelled and most of the light is diffracted into two modes called the -1 and +1 orders which diffract at an angle from the normal axis. By measuring the intensity of either the -1, 0, +1 order modes, the relative separation between the alternating fingers can be monitored.

A diagram of two versions of interdigital cantilever is shown in Figure 4.2.2. The fingers of each cantilever are illuminated from the top with a coherent source and the resulting diffracted orders in reflection are shown schematically. The cantilever is defined such that when a force acts on the tip, only alternating fingers which are connected to the outer, or moving, portion of the cantilever are displaced. The remaining set of fingers, or reference fingers, are attached to the inner portion of the cantilever and remain fixed. Figure 4.2.2a shows a longitudinal interdigital cantilever where the fingers are directed along the cantilever axis. In the transverse version, the fingers are perpendicular to cantilever axis (Figure 4.2.2b). There is little difference between the longitudinal and transverse versions except for the axis of the diffraction pattern, which is perpendicular to the cantilever for longitudinal fingers and parallel to the cantilever axis for transverse fingers. The longitudinal geometry is simpler, but it is

---

1. O. Solgaard, F.S.A Sandejas, and D.M. Bloom, "Deformable grating optical modulator," Optics Letters, vol. 17, no.9, 688 (1992).

2. A. Atalar, S.R. Manalis, S.C. Minne, C.F. Quate, "Interdigital Deflection Sensor for Microcantilevers," Patent pending.

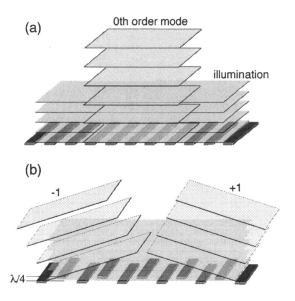

*Figure 4.2.1   Plane waves are (a) reflected toward the source in the 0th order
mode, and (b) diffracted into -1 and +1 order modes when alternating
fingers of the grating are displaced by one quarter wavelength.*

not desirable for cantilever arrays since the higher order diffraction patterns from
neighboring cantilevers will overlap each other. The simulations in this chapter and
the results in the following chapter are obtained for the transverse version. With the
exception of array operation, we have obtained similar results with the longitudinal
and transverse versions.

When the cantilever is illuminated, the fingers form a diffraction grating. The tip displacement can be determined by measuring the intensity of the diffracted modes.
This is depicted schematically in Figure 4.2.3 where the diffraction pattern has been
calculated numerically using a far-field approximation at a distance, h, above a cantilever with a grating period, d. For these simulations, the cantilever was illuminated
with a Gaussian beam with a diameter of 20 μm and a wavelength of λ=670 nm. A
cross-section of the interdigitated fingers is shown below to indicate the cantilever
deflection. In Figure 4.2.3a, the cantilever is not deflected and the dominant reflected
beam is the 0th mode. Also shown are the 2nd order modes created from light dif-

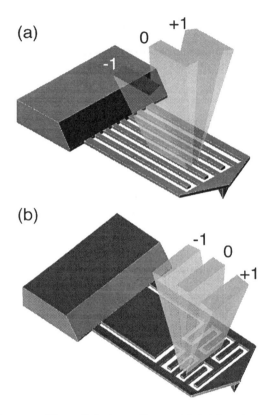

Figure 4.2.2   Diagram of interdigital cantilvers with (a) longitudinal, and (b) transverse fingers.

fracted by a periodic grating with a period of d/2. The angular separation between the 0th and 2nd mode is given by

$$\sin \theta_{0, 2} = \frac{\lambda}{d/2}.$$ (4.2.1)

As the tip is displaced by an external force, the interference between the light reflecting off the reference fingers (dark) and the moving fingers (gray) causes the 0th mode intensity to decrease while a 1st order mode appears. When the cantilever has deflected by $\lambda/4$, where λ is the wavelength of the illumination source, the 0th mode is

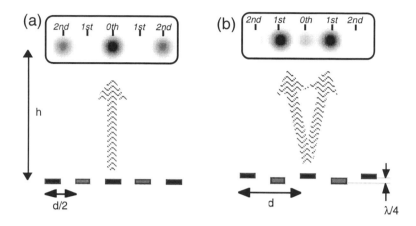

Figure 4.2.3   Calculated diffraction patterns from an interdigital cantilever.

minimized and the 1st mode is maximized (see Figure 4.2.3b). The angular separation between the 0th and 1st mode is given by

$$\sin\theta_{0,1} = \frac{\lambda}{d}. \tag{4.2.2}$$

Assuming $d \gg \lambda$, the separation between the 0th and 1st modes is equal to half of the separation between the 0th and 2nd modes. This is expected since the grating in Figure 4.2.3a has a period that is half as large at the grating in Figure 4.2.3b.

The cantilever deflection can be determined by either measuring the intensity of the 0th mode, 1st mode, or the difference between the modes. The distance between the 0th and 1st mode is equal to h x λ/d and is roughly 2 mm for the parameters in Figure 4.2.3.

A key feature of interdigital detection is that the position of each order remains fixed in space. Only the intensity is changed when the cantilever is deflected. This reduces the alignment requirement of the photodetector when compared with the optical lever sensor. In the case of the optical lever, deflection is determined by measuring the position of the reflected laser spot. This requires the split-photodiode to be precisely aligned.

The intensity of the 0th order diffracted mode can be calculated by summing the light reflected off each set of interdigitated fingers. We assume an incident plane wave in the form $\cos(\omega t + 2\pi z/\lambda)$. The amplitude of the light reflected at $z=0$ from the reference fingers is $\cos(\omega t)$ and from the moving fingers is $\cos(\omega t + 2\pi\delta/\lambda)$, where δ is deflection of the moving portion. Adding the cosine terms, we find that the intensity of the 0th order component is:

$$I_o \sim \cos^2\left(\frac{2\pi\delta}{\lambda}\right). \tag{4.2.3}$$

Therefore, the reflected beams from moving fingers and reference fingers add constructively when δ is equal to 0, $\lambda/2$, λ, $3\lambda/2$... Similarly, the intensity of the first order component is:

$$I_1 \sim \sin^2\left(\frac{2\pi\delta}{\lambda}\right). \tag{4.2.4}$$

The reflected light from moving fingers and reference fingers adds constructively when $\lambda/4$, $3\lambda/4$, $5\lambda/4$, etc.

The phase difference between the two reflected beams is $2\pi\delta/\lambda$, assuming that the incident beam is normal to the cantilever plane. Experimentally, it is difficult to illuminate the cantilever and detect the diffracted orders when the illumination is incident with an angle of zero degrees. If the illumination is incident with a small angle, ϕ, the overall phase difference is reduced by a factor of $\cos\phi$. The incidence angle should be kept as small as possible for maximum sensitivity.

There are two issues that should be considered when designing interdigital cantilevers. First, the spatial separation of the diffracted modes must be larger than the width of a given mode. If the modes are not well separated, they overlap each other and the sensitivity is reduced. The beam width for a mode at the observation plane is approximately $\lambda h/dN$, where N is the number of finger pairs. The separation between the 0th and 1st mode is given by (4.2.2) and is equal to $\lambda h/d$ for $d\gg\lambda$. The ratio of the spatial separation between successive modes to the beam width is simply N. This ratio depends on the number of fingers, but is independent of the observation distance,

h. We have found that for N greater than 2, the modes are sufficiently separated such that they can be independently measured with a photodector.

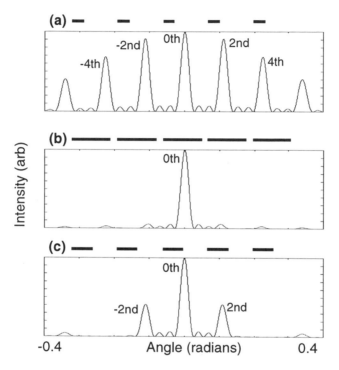

Figure 4.2.4   Optical intensity vs. angle from a one-dimensional grating with varying ratios of finger width, b, to finger spacing, s. (a) b/s=0.2 (b) b/s=5 (c) b/s=1.   The illumination wavelength is 670 nm.

The second design issue is the width of the grating fingers, b, and the finger spacing, s. The intensity distribution from a one-dimensional grating can be calculated theoretically and is given by[3]

$$I = I_o\left(\frac{\sin\beta}{\beta}\right)^2\left(\frac{\sin(2N\beta\gamma)}{2N\sin\gamma}\right)^2,$$ (4.2.5)

3. G. Fowles, *Introduction to Modern Optics*, Holt, Rinehart and Winston, New York, (1967).

where

$$\beta = \frac{\pi w b}{\lambda} \sin\theta \quad,$$ (4.2.6)

and

$$\gamma = \frac{\pi(b+s)}{\lambda} \sin\theta .$$ (4.2.7)

It is interesting to plot (4.2.5) as a function of angle, θ, for three cases. If the finger width is sufficiently small compared to the finger spacing, then $\sin\beta/\beta$ is near unity and the intensity of the first few modes will be approximately the same (see Figure 4.2.4a). This geometry is not desirable because only a small percentage of the incident light will be reflected, and it will be distributed among many modes. In Figure 4.2.4b, the finger width is much larger than the finger spacing, and most of the incident light is reflected specularly. This is the most efficient geometry. Finally, the angular intensity distribution is plotted in Figure 4.2.4c for an equal finger width and spacing.

4.3 Optical Simulations

In order to simulate the intensity distribution of the diffracted modes as a function of cantilever deflection, we need to solve for the diffraction pattern. One approach is to apply the Huygens-Fresnel principle: *a geometrical point source of light will give rise to a spherical wave propagating equally in all directions.*[4] The spherical wave emanating from a point source is described by

$$\psi(x) = \frac{e^{ikx}}{r}$$ (4.3.1)

and represents a solution to the Helmholtz equation,

---

4. H. Weaver, *Application of Discrete and Continuous Fourier Analysis*, John Wiley and Sons, New York (1983).

$$\nabla^2 \psi(x) + k^2 \psi(x) = 0. \tag{4.3.2}$$

Since we have a distribution of points emitting spherical waves and the Helmholtz equation is a linear differential equation, (4.3.1) can be summed over all emitting points. The result is the diffraction equation[5]

$$\Psi_S(x, y, z) = \iint \Psi_R(u, v) K(u, x) \frac{e^{ikr}}{r} du\, dv, \tag{4.3.3}$$

where K(u,x) is called the inclination factor and can generally be ignored provided that the incidence and observation directions are nearly the same. Referring to Figure 4.3.1, this equation accurately describes the propagation of optical radiation from plane R (cantilever plane) to plane S (detector plane) in free space, separated by a distance z.

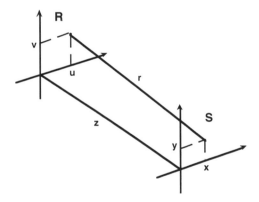

Figure 4.3.1   Coordinate system used in diffraction calculations.

Equation(4.3.3) can be simplified by assuming that the length scale of the object in plane R is small compared to the vertical position of the observation point (z) so that r ~ z. The result is called the Fresnel equation and is given by

$$\Psi_S(x, y, z) = C(x) \iint g(u, v) e^{-2\pi i \frac{(ux + vy)}{\lambda z}} du\, dv \tag{4.3.4}$$

5. J.W. Goodman, *Introduction to Fourier Optics*, 2nd Edition, McGraw Hill (1996).

where

$$C(x) = -\frac{i}{\lambda z} e^{ikz} e^{ik\frac{(x^2 + y^2)}{2z}}$$

(4.3.5)

and

$$g(u, v) = \Psi_R(u, v) e^{ik\frac{(u^2 + v^2)}{2z}}.$$

(4.3.6)

The integral in equation (4.3.4) is the two-dimensional Fourier transform of the function $g(u,v)$. The power received in the plane of the detector is proportional to $|\Psi_S|^2$. Equation (4.3.4) can be solved numerically using MATLAB in three steps:

Step 1. Create a cantilever matrix. Stationary fingers are represented by ones while the movable fingers are set to $e^{4\pi i\delta/\lambda}$ where δ is the vertical separation between the interdigitated fingers. All other matrix elements are set to zero. δ varies along the length of the cantilever and has the value Z_0 at the tip (actual displacement) and 0 where the fixed and movable portions meet.

Step 2. Multiply the matrix created in Step 1 by an illumination profile (i.e. spherical gaussian, cylindrical gaussian, etc.) to form Ψ_R.

Step 3. Multiply Ψ_R by $e^{ik(u^2 + v^2)/2z}$ and take the 2D Fast Fourier Transform (FFT) to form Ψ_S.

It is also possible to find the optical disturbance in the near field region using the angular spectrum approach. This method involves calculating the FFT of Ψ_R, multiplying by $e^{ik_z z}$, and taking the inverse FFT. This is demonstrated in Figure 4.3.2 where the diffraction pattern is calculated for various heights above the interdigital cantilever.

In the far field region where z >> (cantilever width), multiplying by the term $e^{ik_z z}$ has little effect. Therefore the power received at the detector plane can be found by simply taking the FFT of Ψ_R. This is known as the Fraunhofer region.

Calculations in the near field regime require a great deal of computer memory, speed, and time. However they are useful for illustrating how the original intensity distribution of a grating pattern at z=0 evolves to a set of diffracted modes at z ~ 1 cm. For

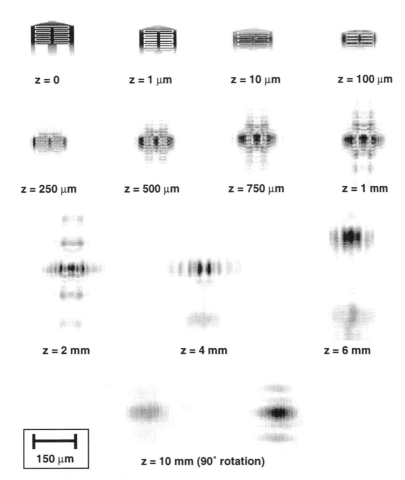

Figure 4.3.2   Simulated diffraction pattern from an interdigital cantilever illuminated by a cylindrical source.

the application of the interdigital cantilever, the far field approximation is generally acceptable since the cantilever dimensions are much smaller than the desired distance to the detector.

Figure 4.3.3a shows a typical interdigital cantilever simulated by the process described in step 1. The cantilever is 150 µm wide and the length ranges between 200 and 500 µm depending on the desired spring constant. Since the cantilevers will ulti-

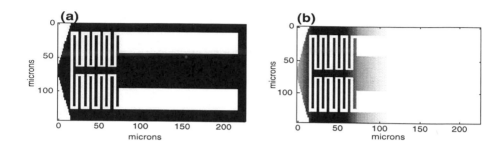

Figure 4.3.3 (a) Initial cantilever matrix. (b) Cantilever matrix representing illumination by a cylindrical Gaussian beam.

mately be integrated into linear arrays, it is convenient to illuminate a set of cantilevers in parallel using a cylindrical Gaussian beam. This is simulated by multiplying each element of the matrix in Figure 4.3.3a by each element of a cylindrical gaussian intensity matrix. The result, $|\psi_R|^2$, is shown in Figure 4.3.3b.

The diffraction pattern at the plane of the detector was calculated using Steps 2 and 3 and is shown in Figure 4.3.4 for a cantilever with a non-zero deflection (the first order mode vanishes for zero deflection). The pattern was determined for a detector position of z=3 cm above a cantilever illuminated by a laser with a wavelength of 620 nm. It would be most desirable for the diffraction grating to be aligned along the length of the cantilever as in Figure 4.2.2a, since this would improve the efficiency by eliminating unwanted optical interference from the interdigitated finger support structure. However, when measuring the deflection of cantilever arrays, such an arrangement would limit the cantilever density since neighboring devices would need to be placed with sufficiently far to achieve neglible interference.

The power delivered to the detectors for the 0th and 1st order modes can be recorded as a function of the net displacement at the end of the cantilever. This is shown in Figure 4.3.5 along with the normalized power difference of the detectors. The power is roughly linear within a range of a thousand angstroms about the steepest portion of a given interference fringe. If the cantilever must detect topography greater than this range, feedback can be used to maintain a constant deflection.

Experimentally, the actual deflection of a single cantilever can be determined by measuring the intensity of the 0th or 1st order modes. If the cantilever is operated at a setpoint where the intensity of the 0th and 1st modes are roughly equal, unwanted

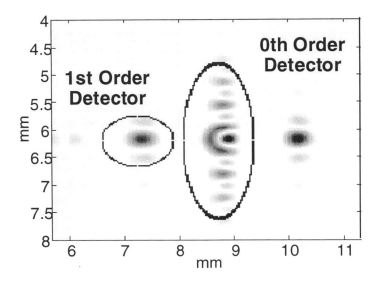

Figure 4.3.4   Diffraction pattern 3cm above the cantilever of Figure 4.3.3

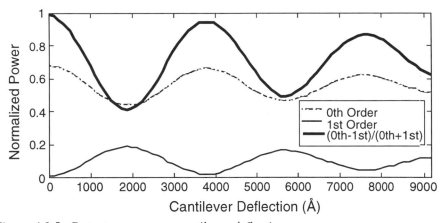

Figure 4.3.5   Detector power vs. cantilever deflection.

intensity fluctuations of the illumination source can be removed by subtracting the 0th and 1st modes. Experimentally, we have found that intensity fluctuations of the laser diode are not the dominant noise source. Assuming that the source is stable, the overall signal is doubled when using the differential measurement, but the laser shot noise is increased by a factor $\sqrt{2}$. As a result, the signal-to-noise ratio (SNR) is only marginally improved through a differential measurement. For one cantilever, a differential measurement adds little complexity over an intensity measurement of a single mode.

If multiple cantilevers are operated in parallel arrays, a crucial parameter is the spacing between cantilevers. The width of the diffracted modes in terms of the cantilever width, w, observation position, h, and illumination wavelength is approximately hλ/w. To avoid optical interference from neighboring cantilevers, the ratio of the cantilever width to the diffracted mode width should be sufficiently greater than unity. This implies a lower limit of the cantilever width and is given by

$$w_{min} \sim \sqrt{\lambda h} \qquad\qquad (4.3.7)$$

This yields a minimum width of ~80 μm for a 1 cm observation point. Solutions to the Fresnel equation indicate that the fist order peak measured at a height of 1 cm has a full-width at half-max (FWHM) of 50 μm for a 150 μm wide cantilever (see Figure 4.3.6). It is therefore possible to place cantilevers in a linear array with a period of 200 μm without optical coupling. Using the superposition principle, a cantilever array can be simulated by adding the complex optical disturbance generated by each cantilever with an offset period of 200 μm (see Figure 4.3.7).

4.4 Minimum Detectable Deflection

The main sources of noise to consider are 1/f, thermal-mechanical noise of the cantilever, intensity and wavelength fluctuations of the laser (also 1/f), shot noise of the detector, resistor Johnson noise, and noise from front-end electronic components. Experimentally, we have found that the dominant noise source for frequencies below 1 kHz is 1/f. We believe that the 1/f noise is likely due to wavelength fluctuations of the illumination source which we analyze in the next chapter. At frequencies above 1 kHz, the noise is independent of frequency and can be estimated by calculating the shot noise and resistor Johnson noise from the photodiode electronics. Below we compare the RMS noise with the detection sensitivity. We also estimate the minimum detectable deflection (MDD) by assuming a signal-to-noise ratio of unity. Finally, we compare the MDD with the thermal mechanical noise of the cantilever.

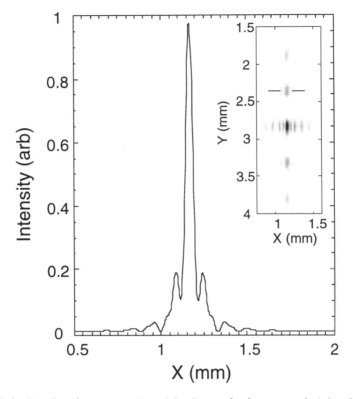

Figure 4.3.6 Simulated cross-section of the first order beam at a height of 1 cm.

The sensitivity, S, of the cantilever in terms of current generated by the photodiode per angstrom of deflection is given by

$$S = \frac{P \Re \Delta P_n}{\Delta x},$$ (4.4.1)

where $\Delta P_n/\Delta x$ is the change in normalized power per cantilever deflection (sensitivity of the diffraction grating as determined by the slope of Figure 4.3.5), $\Re$ is the respon-

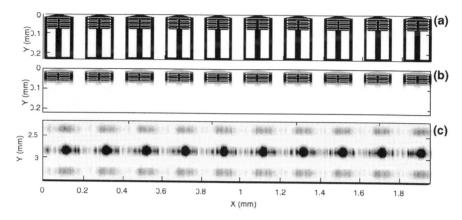

*Figure 4.3.7   (a) Cantilever array. (b) Cantilever array illuminated by a
cylindrical beam. (c) Diffracxtion pattern at a height of 1 cm.*

sivity of the photodiode, and P is the average power incident on the detector.[6] The
responsivity is the amount of photocurrent, I, produced by a given optical power,

$$\Re = \frac{I}{P} = \frac{\eta e}{h \upsilon} = \frac{\eta \lambda}{1.24}, \qquad\qquad (4.4.2)$$

where e is the electronic charge, λ is wavelength of the laser in microns (with photon
energy hv), and η is the quantum efficiency (number of electron-hole pairs created
per incident photon).   Illuminating the cantilever with a 1 mW photodiode at 670 nm
wavelength will yield an incident detector power, P, on the order of 0.1 mW. The
quantum efficiency of a silicon detector at 670 nm is nearly 1, which gives a responsivity of 0.54 A/W. At the steepest portion of the curve in Figure 4.3.5, $\Delta P_n / \Delta x$ is
roughly 4.4×10^{-4} Å^{-1}, which yields a sensitivity of 24 nA/Å.

The noise of the optical detection system is primarily produced by shot noise of the
photodetector and the Johnson noise of the resistor used to convert the photo-current
to a voltage. The shot noise is independent of frequency and arises from the statistics
of photons incident on the photodetector which produce a current,

---

6. S.M. Sze, *Physics of Semiconductor Devices*, 2nd Edition, John Wiley and Sons, New York
(1981).

$$\langle \delta i_{shot}^{2} \rangle = 2e\eta PB, \qquad (4.4.3)$$

where B is the detection bandwidth. For an incident power of 0.1 mW, $\langle \delta i_{shot}^{2} \rangle^{1/2}$ is 6 pA/√Hz. The Johnson noise produced by the resistor R in the trans-impedance amplifier of the photodetector is given by

$$\langle \delta i_{J}^{2} \rangle = \frac{4k_{b}TB}{R}, \qquad (4.4.4)$$

where k_b is the Boltzmann constant and T is the temperature. For a current-to-voltage resistor of 10 kΩ, $\langle \delta i_{J}^{2} \rangle^{1/2}$ is 1 pA/√Hz at room temperature. The total noise is the sum of (4.4.3) and (4.4.4) or

$$\langle \delta i_{T}^{2} \rangle = \sqrt{\frac{4k_{b}TB}{R} + 2e\eta PB}, \qquad (4.4.5)$$

which is 6 pA/√Hz and is dominated by the shot noise. In the case of a single interdigital cantilever, the diffracted power is typically large enough such that the total noise is dominated by shot noise and the resistor Johnson noise can be ignored. However, for cantilever arrays the diffracted power is generally lower and the Johnson noise should be considered.

Assuming a signal-to-noise ratio of unity, the MDD of the cantilever is determined by dividing the total current noise given in (4.4.5) by the sensitivity given in (4.4.1) to obtain

$$MDD = \frac{\sqrt{\frac{4k_{b}TB}{R} + 2e\eta PB}}{P\Re\frac{\Delta P_{n}}{\Delta x}}. \qquad (4.4.6)$$

The minimum detectable deflection of the cantilever measured with a silicon photodiode is approximately 3.2×10^{-4} Å/√Hz. (4.4.6) indicates that although the shot noise increases with square root of P, the sensitivity is proportional to P. The MDD can therefore be improved by increasing the incident laser power.

Mechanical simulations of an interdigital cantilever with a 1 μm thickness yield a resonant frequency near 30 kHz and a spring constant of 0.1 N/m. The Q of the cantilever typically ranges between 50 and 100 in air. We can estimate the thermal-

---

Cantilevers with Interdigital Deflection Sensors　　　　　　　　　　**97** mechanical noise of the cantilever by approximating it as a simple harmonic oscillator.

A classical one-dimensional harmonic oscillator is described by

$$m\frac{d^2z}{dt^2} + \frac{k}{Q\omega_o}\frac{dz}{dt} + kz = 0 \qquad (4.4.7)$$

where z is displacement, k is the spring constant, and Q is the quality factor. Using (4.4.7) it can be shown that thermal mechanical fluctuation in position of the oscillator is[7]

$$\langle\delta z^2\rangle^{1/2} = \sqrt{\left(\frac{4k_BTB}{k\omega_o}\right)\frac{Q}{[Q^2(1-\omega^2/\omega_o^2)^2 + \omega^2/\omega_o^2]}} \cdot \qquad (4.4.8)$$

If the frequency range of the detection is well below the mechanical resonance of the system, ω_o, (4.4.8) can be approximated by

$$\langle\delta z^2\rangle^{1/2} = \sqrt{\frac{4k_BTB}{k\omega_o}} \qquad (4.4.9)$$

Using this simple approximation for the cantilever, we estimate its thermal mechanical noise at room temperature as 3.2×10^{-3} Å/√Hz. Therefore the thermal mechanical noise of the cantilever will limit the resolution even though the optical detection system can measure deflections of 3.2×10^{-4} Å/√Hz.

For many applications, the cantilever can be stiffer than 0.1 N/m. For example, if the thickness is changed to 4 μm, the spring constant will increase to 1-10 N/m. This will yield a thermal mechanical noise of approximately 1.5×10^{-5} Å/√Hz which is below the noise level of the detection system.

---

7. D. Sarid, *Scanning Force Microscopy*, Oxford University Press, New York (1991).

CHAPTER 5

Operation of the Interdigital Cantilever

5.1 Microscope Description

The interdigital cantilever introduced in Chapter 4 can be operated with a commercial AFM that is configured for the optical lever without modification. This is because the laser focal spot is generally large enough to illuminate a few sets of interdigital fingers and the split photodetector is typically placed more than one centimeter from the cantilever which is in the far field diffraction region.

Our system consists of an unmodified commercially available optical lever microscope head[1] combined with a homebuilt base, scanner, and electronics. This approach allows us to benefit from a compact optical head that is well designed while maintaining operating flexibility through analog control and detection circuitry. A block diagram of the microscope is shown in Figure 5.1.1.

As an illustrative example, if a 670 nm wavelength laser is used to illuminate an interdigital cantilever with a 12 μm grating pitch and the split photodetector is placed 4 cm from the cantilever, then the 0th and 1st modes will be separated by roughly 2 mm. Either the intensity of the 0th mode, 1st mode, or difference between the two modes can be measured by the split photodiode.

---

1. Multimode Head, Digital Instruments, Santa Barbara, CA.

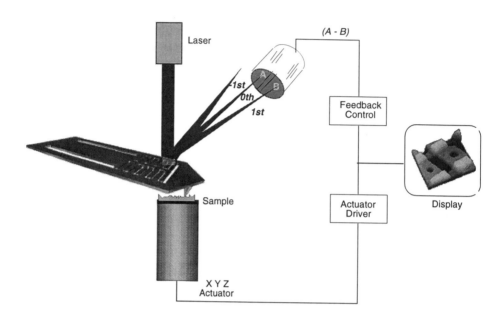

Figure 5.1.1   Block diagram of a standard atomic force microscope equipped with an interdigital cantilever.

We choose to work primarily with the transverse design where the interdigitated fingers are perpendicular to the cantilever length. The modes of the resulting diffraction pattern are then deflected along a line that is parallel to the cantilever length and will minimize optical interference when adjacent cantilevers are present. Figure 5.1.2 shows a typical force curve that is obtained by measuring the intensity of the 0th, 1st, and difference between the 0th and 1st modes as a function of cantilever deflection. The cantilever is deflected by applying a force at the tip with a piezo tube. A laser diode with a wavelength of 670 nm is focused to a ~20 μm spot and aligned to a set of interdigitated fingers. The spot is placed such that the longitudinal finger support is not illuminated. The intensity of the diffracted modes is measured with a split photodiode placed roughly 4 cm from the cantilever. Note that the amplitude of the interference fringes decreases as the cantilever deflection is increased. This occurs because the plane of the deflected set of interdigitated fingers does not remain parallel to the plane of the reference set when the cantilever bends. In most situations, the interdigital cantilever can be operated within small deflections from equilibrium.

Similar force curves can also be obtained from the longitudinal design by using the lateral cells of a four quadrant photodiode or by rotating the split photodiode by 90 degrees.

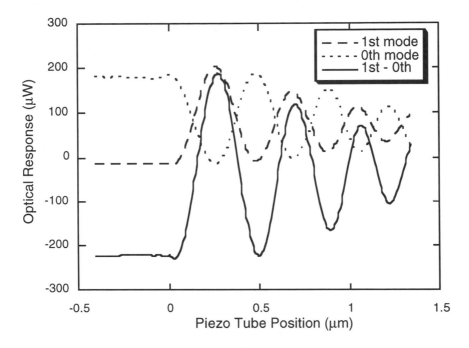

Figure 5.1.2   Force curve obtained by recording the intesity of the 0th and 1st order diffracted modes as the cantilever is deflected by displacing the tip with a piezo tube

The period of the optical response shown in Figure 5.1.2 is 470 nm, which is larger than the expected value of $\lambda/2=335$ nm. The discrepancy can be accounted for by considering the following effects. First, since the force is applied at the tip and the laser beam is focused on the diffraction grating, the actual vertical finger displacement is less than the tip deflection due to the bending of the cantilever. Second, the expected period is slightly increased because the laser beam does not reflect off the diffraction grating with normal incidence.

5.2 Imaging

When imaging in the constant height mode, the tip is placed into contact with a sample and the vertical position of the piezo tube is adjusted to maximize the cantilever sensitivity. The force curve in Figure 5.1.2 indicates that the detector output will be linear with the cantilever deflection provided that displacements are less than ~100 nm. However, this range can be increased at the expense of vertical resolution by illuminating fingers that are closer to the cantilever base since the vertical displacement of these fingers is reduced. Thus, translating the illumination spot along the cantilever length provides a simple means of controlling the dynamic range of the interdigital sensor.

Atomic images of graphite taken with the interdigital cantilever are shown in Figure 5.2.1. These images are acquired with PC compatible analog input/output hardware controlled by LabView software. The analog hardware controls the piezo scanner while sampling data from the output of the photodiode preamplifier. The data in Figure 5.2.1b is unprocessed with the exception of the lower left-hand corner which has been filtered to clearly show the hexagonal structure of graphite. The square-like quality of the atomic image results from a relatively large contact force between the tip and sample. To maximize sensitivity, the outer portion of the cantilever must be deflected, or biased, by applying a force at the tip. This is undesirable because it is not possible to arbitrarily choose the tip-to-sample force while maintaining maximal sensitivity. In Figure 5.2.1, the sensitivity was maximized by deflecting the cantilever by roughly 175 nm (see Figure 5.1.2). As a result, we estimate that the tip/sample force was 0.4 μN, which is several of orders of magnitude larger than forces typically applied by the AFM. We estimate the spring constant of our cantilevers to be about 2 N/m. When displaced by 175 nm, the resulting force is 0.4 μN. Commercially available Si_3N_4 cantilevers have force constants below 0.01 N/m and can be operated with minimal setpoint deflection.

The linear range of the interdigital cantilever can be significantly expanded by operating in the constant force mode. In this mode, the force setpoint of the feedback (see Figure 5.1.1) is adjusted to maximize the sensitivity. When the cantilever is deflected by the sample topography, the piezo tube compensates for this displacement so that the deflection of the cantilever remains constant. An image of a two-dimensional diffraction grating acquired in this mode is shown in Figure 5.2.2. The vertical position of the piezo tube corresponds to the sample topography and is plotted in Figure 5.2.2a. The output of the photodetector representing the error of the feedback loop is plotted in Figure 5.2.2b. In this imaging mode, the vertical range is limited by

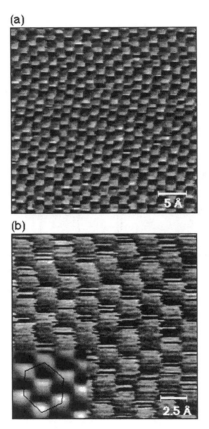

Figure 5.2.1   (a) 35 by 35 A atomic image of graphite obtained with an interdigital cantilever. (b) Image area reduced to 20 by 20 A. The data in the bottom left-hand corner has been filtered to clearly show the hexagonal structure of graphite.

the maximum vertical displacement of the piezo tube which is typically several microns.

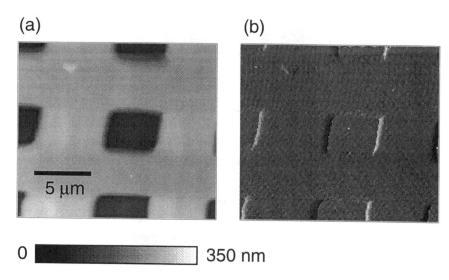

Figure 5.2.2   Image of a two-dimensional diffraction grating acquired in the constant force mode with an interdigital cantilever. (a) Topographic image corresponding to vertical piezo tube position. (b) Error signal from photodiode output.

5.3 Biasing

In order to maximize sensitivity at reasonably low tip-to-sample forces, it is desirable to bias (or deflect) the cantilever. There are several different methods for biasing the interdigital cantilever. The most direct way is to etch the fingers on either the reference or moving portion of the cantilever by one eighth of the illumination wavelength. A vertical offset of $\lambda/8$ in the thickness of alternating fingers will shift the optical response shown in Figure 5.1.2 such that the maximum slope, or sensitivity, will occur at the cantilever equilibrium position.   Since the cantilever is illuminated from the back, the etch must be applied to the side of the fingers that are opposite to the tip. This is very difficult to implement into the fabrication process because only the tip side of the cantilever can be patterned before the cantilevers are released. Once released, the cantilevers are too fragile and the topography of the back side is too large to apply standard processing techniques.

However, the concept of biasing by etching the back of the interdigitated fingers can be demonstrated by using a focused ion beam (FIB). The FIB has the capability of imaging an object in a similar fashion to the scanning electron microscope (SEM) along with the ability to physically alter many types of materials through ion bombardment. It can be used to precisely remove a layer of silicon off the reference set of fingers that is roughly $\lambda/8$ thick. A SEM of the resulting cantilever is shown in Figure 5.3.1. In this image, the reference fingers appear to be darker than the moving

Figure 5.3.1   SEM micrograph of an interdigital cantilever that has been biased by focused ion beam milling.

fingers and the step height between the fingers and the longitudinal support is ~ 80 nm. The resulting force curve from this cantilever, shown in Figure 5.3.2, reveals a horizontal shift in the optical response thereby indicating that the cantilever is sensitive at the equilibrium position.

Although the FIB is a precise and direct method of etching silicon, it is inherently a serial process and it takes about an hour to bias a single cantilever. Cantilevers can be biased in parallel by depositing and patterning a dissimilar material on the longitudinal support for the reference fingers. The natural stress between this material and the silicon will induce bending and cause the reference fingers to be elevated from the

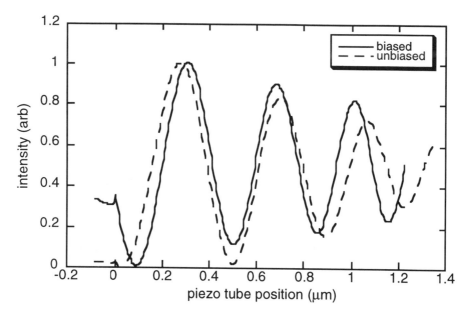

Figure 5.3.2   Force curve from a biased (solid line) and unbiased (dashed line) interdigital cantilever.

plane of the moving set of fingers. Although this technique can be applied to an entire wafer in parallel, it can be difficult to control the biasing uniformity since the degree of stress is critically dependent on thickness of the silicon cantilever and the deposited material. Interdigital cantilevers have been biased by evaporating aluminum on the front after processing and the resulting force curve is shown in Figure 5.5.6 on page 116. We speculate that the most reliable results from this method can be obtained by growing thermal oxide on the reference support since the thickness of oxide films can be accurately controlled.

A third method for biasing is to construct the interdigitated fingers with silicon nitride rather than silicon. The front of the alternating fingers can be etched using standard processing techniques and coated with metal. Since silicon nitride is optically transparent, the majority of the incident light will travel through the nitride and reflect off the metal/nitride interface. Although this concept has not yet been demonstrated, we have shown that cantilevers can be fabricated with silicon and silicon nitride regions. The advantage of using a hybrid process rather than an entire nitride process is that high quality tips can be constructed from the silicon and if additional processing is required such as ZnO actuators, it can be applied to the front side.

5.4 Resolution and Frequency Response

The resolution of the interdigital cantilever is estimated by measuring the noise spectral density of a free standing cantilever (see Figure 5.4.1). To calibrate this measure-

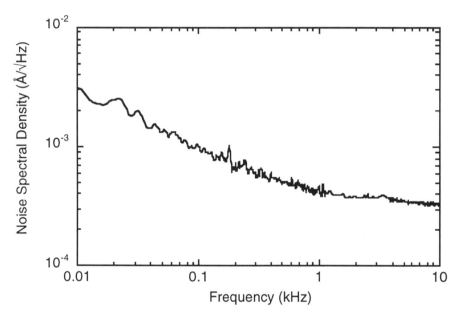

Figure 5.4.1   Noise spectral density of a biased interdigital cantilever. The maximum slope from the force curve shown in Figure 5.3.2 was used to calibrate the noise spectral density.

ment in terms of position, the noise spectral density is divided by the cantilever sensitivity, which is the maximum slope of the force curve shown in Figure 5.3.2. This yields an RMS noise of about 0.02 Å in a 10 Hz to 1 kHz bandwidth.

The interdigital cantilever is analogous to the fiber optic interferometer developed by Rugar[2]. Rugar points out that small changes in the laser wavelength or phase will change the detected optical power in a similar fashion to cantilever deflections. If the fiber and reflector are separated by a distance d and the wavelength is shifted by $\Delta\lambda$ the resulting phase difference is

$$\Delta\varphi = 2\pi\left(\frac{d}{\lambda + \Delta\lambda} - \frac{d}{\lambda}\right) \tag{5.4.1}$$

If the arms are displaced by an additional distance Δd while the wavelength is held constant, a similar phase shift of

$$\Delta\varphi = 2\pi\left(\frac{d + \Delta\lambda}{\lambda} - \frac{d}{\lambda}\right) \tag{5.4.2}$$

is created. (5.4.1) and (5.4.2) indicate that fluctuations in position are related to fluctuations in wavelength by

$$\Delta d = d\left(\frac{\Delta\lambda}{\lambda}\right) \tag{5.4.3}$$

(5.4.3) shows that the RMS noise of an interferometer resulting from wavelength fluctuations of the illumination source increases linearly with fiber-to-reflector spacing. Experimentally, this has been measured by Rugar, and the data is plotted in Figure 5.4.2.

Figure 5.4.2 indicates that the RMS noise for a near zero fiber-to-reflector spacing is ~0.04 Å. Since the interdigital cantilever used to obtain our noise measurement of 0.02 Å is biased with a 80 nm spacing, we conclude that the 1/f noise in the low frequency band is likely due to wavelength or phase fluctuations of the laser diode. In the high frequency band (above ~ 1 kHz), the noise spectrum is flat and the resolution is primarily limited by shot noise.

The mechanical frequency response of the interdigital cantilever can be analyzed by using a thin slab of piezoelectric material to drive the cantilever along with the substrate. The mass of the 2 mm x 3 mm x 0.5 mm die on which the interdigital cantile-

2. D. Rugar, H.J. Mamin, and P. Guethner, "Improved fiber-optic interferometer for atomic force microscopy," Appl. Phys. Lett. vol. 55, 2588 (1989).

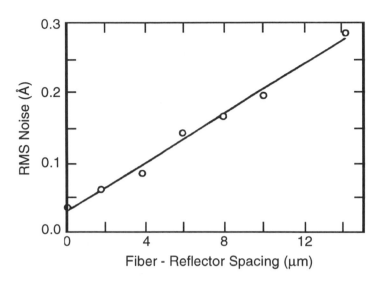

Figure 5.4.2   Measured noise in a 1Hz to 1kHz bandwidth as a function of fiber-to-reflector spacing.   From reference 2.

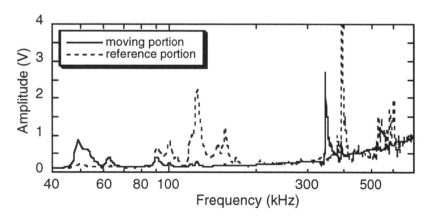

Figure 5.4.3   Frequency response of the reference and moving parts of an interdigital cantilever as measured by an optical lever.

ver is fabricated is low and it is common to use this technique to drive the cantilever at frequencies in excess of 500 kHz. Since the interdigital cantilever consists of a reference beam and a moving beam that are connected to a fixed substrate, it is helpful to study each portion independently. This can most readily be accomplished by using optical lever detection where the incident laser beam is focused either on the triangular part of the moving portion or on the longitudinal support of the reference portion. The output of the split photodiode is connected to an RMS to DC converter and is plotted versus drive frequency to the piezoelectric slab. Figure 5.4.3 reveals that the fundamental mode of the moving portion is at 50 kHz while the reference portion, which is shorter and contains less mass, resonates at a higher frequency near 100 kHz. The next higher order mode of the moving and reference portion resonates at 350 kHz and 400 kHz, respectively. In this measurement, the tip end of the cantilever is free.

5.5 Interdigital Cantilever Arrays

Operation of interdigital cantilever arrays without position feedback is limited for most imaging applications because the detection output is nonlinear and each cantilever must be biased for sensitivity. However, provided that the variation of the sample surface and the height of the tips are less than $\sim\lambda/16$, all cantilevers within an array can be biased as a unit by using the tip-to-sample force to bend the moving portion by $\lambda/8$. It is then possible to image a surface in the constant height mode as long as the topography falls within the linear range. If the illumination source is aligned to the fingers nearest to the base of our cantilevers, the linear range can be over 200 nm.

Parallel operation in the constant height mode with multiple interdigital cantilevers can be accomplished with a cylindrical illumination source and a photodetector array. This concept is shown schematically in Figure 5.5.1. In this drawing, a laser beam is cylindrically focused on the array so that the interdigitated fingers of each cantilever are illuminated. The diffracted orders from a given cantilever are produced on a line oriented along the cantilever length. The photodetectors are aligned as a unit to the desired mode. Since both the longitudinal finger support of the reference and moving portions of the cantilevers is also illuminated, diffracted modes will also be created in the direction of the array (horizontal axis). These unwanted modes will be primarily adjacent to the 0th mode. By measuring the intensity of the +1 or -1 orders rather than the 0th order, optical coupling from the horizontal diffracted orders can be minimized. The deflection of the cantilevers can be measured by positioning a photodetector array with a horizontal pitch equal to the cantilever spacing. The vertical size of each detector should be sufficiently larger than the spot size of a given order but

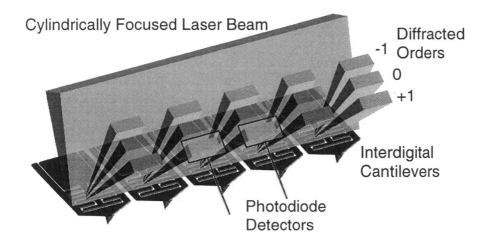

Figure 5.5.1   Illustrative schematic of parallel operation of interdigital cantilevers.

small enough so that the neighboring modes are not detected. Since the deflection is determined only through intensity, the vertical alignment is not crucial as long as the diffracted modes are incident on the detector.

An SEM micrograph of an array of interdigital cantilevers is shown in Figure 5.5.2. The array contains integrated tips and was fabricated using the process described in Chapter 9. Cantilevers range in length from 220 to 400 μm, are 2 μm thick, and are spaced on a 200 μm pitch.

Parallel operation with two interdigital cantilevers can be demonstrated by using a split photodiode detector to measure the deflection of any two consecutive cantilevers within an array. A schematic of the microscope head used to measure the diffracted mode intensity from interdigital arrays is shown in Figure 5.5.3. This microscope differs from the microscope used to measure a single interdigital cantilever (see Figure 5.1.1) in two ways. First, an imaging lens is used to project the diffracted orders from a plane located a few millimeters above the cantilevers to the photodetector plane. The lens has a focal length of 1.2 cm which creates a 1:1 image at twice this length. This lens is necessary because the optical path from the cantilevers to the photodetectors is approximately 10 cm and the far field diffraction pattern from a single cantilever appears at a distance of only a few millimeters. If the lens were absent,

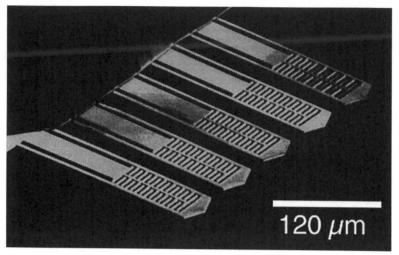

120 μm

Figure 5.5.2 SEM micrograph of an array of interdigital cantilevers. The integrated, single-crystal silicon tips are too small to be seen in this view.

the divergence of the diffracted orders from one cantilever would optically interfere with orders from adjacent cantilevers. In other words, the lens images the diffracted light from the array at a distance that is in the far field limit of a single cantilever but in the near field regime from the array as a whole.

A second difference from the single cantilever setup is that a cylindrical lens is used to focus the laser diode beam. A ~1 mm wide slit is placed before this lens to limit the width of the focused spot. If the beam is focused to a spot that is too narrow, the diffracted orders from a given cantilever will diverge rapidly and they will interfere with each other in the far field limit. The power of the 630 nm laser diode is held constant at a few milliwatts and a commercially available split photodiode is used to measure a single order from two consecutive cantilevers. Since the vertical height of the photo-detectors is several millimeters, an aperture is used to block all modes except for the +1st order.

An optical micrograph of an interdigital cantilever array positioned above a memory portion of an integrated circuit chip is shown in Figure 5.5.4. The optical image is focused on the array and the bright areas of the longitudinal finger support structure

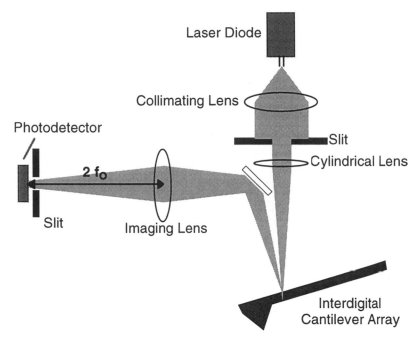

*Figure 5.5.3   Schematic of the microscope head used to measure the deflection of
interdigital cantilever arrays. For simplicity, only the first order
diffracted modes are shown. The slit adjacent to the photodetector
blocks the other orders.*

indicate scattered light from the cylindrically focused laser diode. The interdigitated
fingers are not resolved at this magnification.

Once the array is illuminated, the photodetectors are temporarily removed and the
microscope optics are tested by using an optical microscope to focus on the photodetector plane. This allows a selected set of diffracted modes from all cantilevers to be
inspected. The intensity of these modes can be modulated by engaging the array with
a sample surface. Since the working distance of our microscope is quite large, only a
single mode from each cantilever is incident on the objective lens. An optical image
focused on the photodetector plane of the first order modes of an array containing five
cantilevers is shown in Figure 5.5.5a. In this image, there are five spots spaced by 200
μm corresponding to cantilevers 1-5 shown in Figure 5.5.4. Each spot contains two
overlapping lobes which represent the first order modes emanating from the grating

*Figure 5.5.4 Optical micrograph of an interdigital cantilever array positioned
above a portion of an integrated circuit memory. The bright areas on
the longitudinal support for the interdigitated fingers are scattered
light from the cylindrically focused illumination. The magnification
in this view is not high enough to resolve the individual fingers.*

on either side of the longitudinal finger support. The cantilevers in this array have
been slightly biased due to the stress from a thin film of aluminum deposited on the
front side. Even though the cantilevers have not yet been deflected by the sample surface, the intensity of the first order modes is not zero as in the case for unbiased cantilevers. A piezo tube is then used to translate the sample towards the array. In
Figure 5.5.5b, cantilevers 1 and 2 have been deflected by the sample and the first
order mode intensity is altered. In Figure 5.5.5c, cantilevers 3 and 4 are deflected
while the deflection of cantilever 1 and 2 is increased. As the sample continues to be
translated toward the array, the remaining cantilevers are deflected and images at arbi-
trary positions are shown in Figure 5.5.5d and e. For this series of measurements, the
array was intentionally tilted with respect to the sample surface so that the cantilevers
could be selectively deflected. These images indicate that modes from adjacent cantilevers appear to be optically isolated and do not couple as the cantilevers are
deflected.

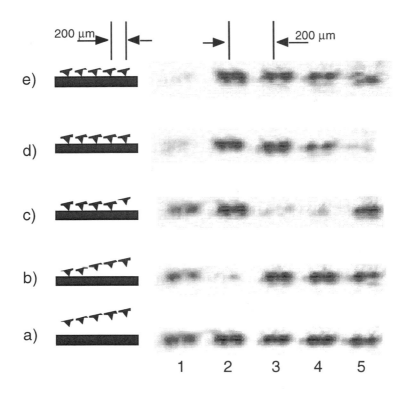

Figure 5.5.5 *First order diffracted modes from an array of five cantilevers imaged
with an optical microscope at the photodetector plane. The far left
column illustrates the position of the array relative to the sample.*

Parallel images in the constant height mode are obtained by using the split photodiode
to measure the 1st order diffracted mode from two cantilevers within the array. The
aperture in front of the photodetectors can be used to block the orders from neighboring cantilevers. Alternatively, a scribe can be used to physically remove all but two
cantilevers from the die. Next, the cantilever array is levelled to the sample surface.
This task is most easily accomplished by monitoring the force curve (see
Figure 5.5.6) from both cantilevers and adjusting the tilt until the contact point is
aligned. It is also possible to align the split photodiode while monitoring the force
curve. Next, the cylindrically focused illumination is adjusted so that only the interdigitated fingers nearest to the cantilever base are illuminated. This increases the

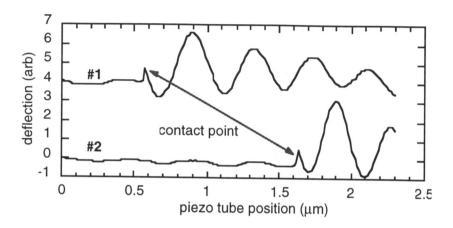

Figure 5.5.6   *Parallel force curves obtained from two interdigital cantilevers. Force curves are monitored before imaging in order to remove the tilt between the array and the sample by aligning the contact point. Note that the fringes before the contact point are due to optical interference from light reflecting off the sample surface.*

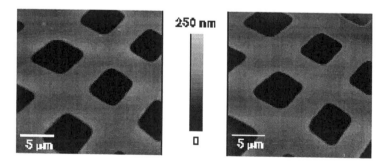

Figure 5.5.7   *Parallel images of a two-dimensional diffraction grating taken in the constant height mode with two interdigital cantilevers spaced by 200 um.*

period of the curves shown in Figure 5.5.6 which maximizes the vertical deflection range where the detector output is linear. The piezo tube is used to extend the sample towards the cantilevers until the center of the linear region is reached. The output of each photodiode is recorded while the piezo tube scans the sample in a raster pattern. Images of a two-dimensional diffraction grating acquired simultaneously by two interdigital cantilevers is shown in Figure 5.5.7. The maximum topographical variation in this image is ~ 230 nm which extends slightly beyond the linear range of the sensors. The vertical resolution of each cantilever is approximately 1 Å in a 1 kHz bandwidth.

5.6 Summary

The relative strengths and weaknesses of the optical lever, fiber optic interferometer, piezoresistor, and interdigital detection schemes are summarized in Figure 5.6.1 in terms of sensitivity, sensitivity dependence on cantilever geometry, alignment requirements, compatibility with integrated actuators, susceptibility to electrical coupling, linearity, and biasing.

	Optical Lever	Fiber Optic Interferometer	Piezoresistor	Interdigital
Sensitivity (1 kHz band)	0.2 A	0.02 A	0.2 A	0.02 A
Sensitivity vs Cantilever Geometry	length	none	length, thickness	none
Alignment Requirements	laser, photodiode (precise)	fiber (precise)	none	laser, photodiode (coarse)
Integrated Actuator Compatibility	requires correction	no	yes	yes
Electrical Coupling Sucseptibility	no	no	yes	no
Linearity	yes	no	yes	no
Biasing	no	yes	no	yes

Figure 5.6.1   Comparison of common deflection sensors.

For single cantilever operation, the optical lever is by far the most commonly used deflection sensor in both industry and the research environment. This is primarily due to its simplicity combined with its ability to routinely achieve atomic resolution. The optical lever is also linear and does not need to be biased. Although the interdigital cantilever is nonlinear and must be biased, it offers four advantages over the optical lever: First, interdigital detection is approximately an order of magnitude more sensitive than the optical levers. The increased sensitivity of interdigital detection is primarily due to its immunity to laser pointing noise and mechanical vibrations of the photodetectors. Second, the alignment of the photodetector is less crucial for the interdigital cantilever since laser beam intensity rather than position is measured. Third, the sensitivity of interdigital detection does not depend on the cantilever length. Finally, the interdigital cantilever combined with an integrated actuator would not require real time correction as in the case of the optical lever. This is possible since the actuator can move both the reference and moving portions as a unit and this does not change the intensity of the diffracted orders.

A crucial aspect of cantilever array systems is the method that is used to monitor the deflection of each cantilever. One of the most well developed techniques for array detection is the piezoresistive sensor. Piezoresistive sensors are attractive because they do not require any external alignment and are linear over a large vertical deflection range. The disadvantage of piezoresistors is that the electronics used to measure the resistance must detect voltage changes down to microvolt levels. This is generally not a problem for single cantilever systems. For parallel arrays, it often desirable to individually control the deflection of each cantilever by integrating electrical actuators that require voltage swings greater than ten volts. In many cases, proper shielding between the sensor and actuator can nearly eliminate electrical coupling for imaging bandwidths of a few tens of kilohertz. However, this task becomes increasingly difficult as the bandwidth is expanded to accommodate high scan speeds. An additional disadvantage of the piezoresistor is that an electrical contact is required for each cantilever. By eliminating this contact, the cantilever array density can be substantially increased.

The drawbacks of piezoresistive sensors can be eliminated by using optical methods to monitor the deflection of cantilever arrays. The penalty for this is the complication of aligning external optical components to the cantilever array. The interdigital detection is an optical based system that requires minimal alignment requirements while achieving high sensitivity.

CHAPTER 6

Cantilever Arrays

6.1 Automated, parallel, high-speed AFM[1]

Microelectronics for the past thirty years has progressed on the single idea of shrinking the feature size on the silicon microchip. G. M. Whitesides tells us that "making things smaller brings benefits — they're less expensive, you get more portability and more performance per dollar."[2] As this trend continues, microelectronics will give way to nanoelectronics. For this to happen we must find methods suitable for dealing with nanoscale devices. The scanning probes are natural tools for this task provided we can increase the size of the scanned area and scan this large area in a reasonable time. We must do all this without sacrificing the resolution and precision of the probe based systems.

The direct approach to this problem is an array of parallel cantilevers. The scanned area of the microscope increases in direct proportion to the number of tips in an array. Integrating deflection sensors onto the cantilever simplifies the operation of the parallel system since it reduces the problems of alignment and maintenance. Many groups

1. Part of this section is reprinted with permission from Minne, *et al.*, "Automated parallel high-speed atomic force microscopy, *Applied Physics Letters* **72**, 2340 (1998). Copyright 1998 American Institute of Physics.

2. "Nanotechnology - Art of the Possible" in *Technology Review*, p85-87, Nov/Dec 1998.

have adopted the piezoresistor for the integrated sensor. We have used this sensor in parallel arrays and used them for both imaging and lithography.[3] Lutwyche et al.[4] have fabricated and demonstrated 2-D arrays of piezoresistive cantilevers.

Parallel piezoresistive cantilevers have also been fabricated by Chui et al.[5] and Ried et al.[6] and applied to systems for data storage. Parallel optical lever cantilever sensors have been demonstrated by Lang et. al.[7] for use in imaging and other sensing applications.

In this section we describe an automated cantilever array design which allows parallel constant force imaging at high speeds. The footprint of the cantilever structure has been modified to occupy a slice only 200um in width, allowing the devices to be placed in a 1-D expandable parallel array. Improved electrical performance of the device allows us to use minimal non-synchronous electronics, permitting simple fabrication of the control system onto a personal computer (PC) expansion card. The integrated electronics, coupled with the PC control, provides automated operation for the array.

Previously in Chapter 3 it was shown that the scan speed of the AFM could be increased by an order of magnitude by integrating a thin film of ZnO on the base of a piezoresistive cantilever. The voltage applied to the ZnO film bends the cantilever to conform to the sample topography and thus maintain constant force on the tip. The sample force is detected by measuring resistance changes in the piezoresistor. In that study, electrical coupling between the ZnO actuator and the piezoresistive sensor required the use of a synchronous system. Although this technique significantly increased the imaging speed, unwanted capacitance in the detection circuitry limited the bandwidth to 6 kHz. Ideally, the imaging bandwidth should be limited by the mechanical resonance of the cantilever when the tip is in contact with a surface. Ultimately the design must have the versatility to control the electrical interaction

3. S. C. Minne, Ph. Flueckiger, H. T. Soh, C. F. Quate, J. Vac. Sci. Technol. B 13, 1380 (1995).

4. M. Lutwyche, C. Andreoli, G. Binnig, J. Brugger, U. Drechsler, W. Haeberle, H. Rohrer, H. Rothuizen, and P. Vettiger, Proceedings IEEE Int'l Workshop on Microelectro Mechanical Systems (MEMS 98), Heidelberg, Germany, Jan 25-29, 1998.

5. B. W. Chui, T. D. Stowe, T. W. Kenny, H. J. Mamin, B. D. Terris, and D. Rugar, Appl. Phys. Lett. 69 2767 (1996).

6. R. P. Ried, H. J. Mamin, B. D. Terris, L. S. Fan, and D. Rugar, J. Microelectromechanical Sys. 6, 294 (1997)

7. H. P. Lang, R. Berger, C. Andreoli, J. Brugger, M. Despont, P. Vettiger, Ch. Gerber, J. K. Gimzewski, J. P. Ramseyer, E. Meyer, and H. J. Guntherodt, Appl. Phys. Lett., 73, 383 (1998).

between the tip and sample (for lithography, modification, etc.). Therefore a provision to bias the tip without interfering with the sensing and actuating properties of the device must be included in the design.

Our cantilever, designed to address these issues, is schematically shown along with a brief process flow in Figure 6.1.1. Inspection of Figure 6.1.1 reveals several key features in the new design. First, the bottom electrode of the actuator (between the ZnO

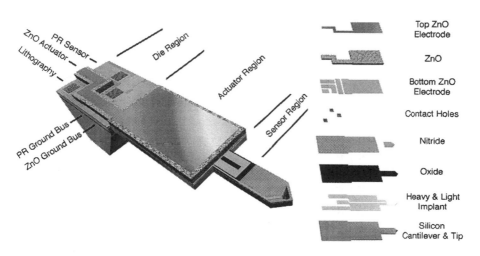

Figure 6.1.1   *Schematic illustration of the expandable cantilever design with integrated piezoresistive sensor and integrated ZnO actuator. The perspective view on the left shows the final device, while the exploded view on the right shows the pattern of all the films in the device.*

and the silicon substrate) serves as a ground plane. This prevents the signals that are applied to the top ZnO electrode from coupling capacitively to the piezoresistor through the silicon substrate. Second, the piezoresistor is defined by a patterned implant and this allows the cantilever to be actuated with a single pad of ZnO. In the previous design, the piezoresistor was defined by physical etching thereby requiring the ZnO to be patterned on both of the cantilever legs. We found that the dual leg ZnO configuration reduced the overall yield of working devices. Third, grounding busses are added for common piezoresistor and common ZnO electrodes. This minimized the tip to tip spacing by eliminating two electrical contacts per cantilever. Finally, a dedicated voltage connection is made to the tip with a heavy implant so tip-to-sample biases can be applied independently of the sensor or actuator signals. A complete description the old design's process, which was followed for this process, is given in Chapter 8.

Micrographs showing the array of cantilevers with the corresponding bonding pad layout are shown in Figure 6.1.2. We have fabricated an array of 50 cantilevers that are spaced by 200 μm yielding an array spanning one centimeter. Figure 6.1.2a shows the 1 cm array next to a U.S. dime for scale. Figure 6.1.2b shows a detail of five cantilevers within the array. The piezoresistor and ZnO regions are clearly seen. The two horizontal metal lines running across the device are the piezoresistor and the ZnO ground busses. The three contacts per cantilever (piezoresistor sensor, ZnO actuator, & tip bias) run vertically down the die to the bonding pads shown in Figure 6.1.2d. Each device occupies a horizontal footprint of 200 μm which is the minimum achievable tip to tip spacing due to the design limitations on the bonding pads. (The bonding services currently available to us can bond two rows of pads where each row is on a 100 μm period. This was the basis for our contact configuration as shown in Figure 6.1.2d.) A close up of the integrated tip is shown in Figure 6.1.2c. The radius of curvature of these single crystal silicon tips is generally less than 100Å. This tip was engineered using gas variations in a plasma etch to produce a high aspect ratio profile that is useful for both imaging and lithography. The tips are covered with a thin film of titanium to enhance the capability for lithography.

For small arrays of cantilevers it is possible to manually control all aspects of the cantilever's operation during an imaging or lithography experiment. However, for massively parallel operations, or operation in a scalable system, it is important to have a compact electronic system for automated control. In the piezoresistive/ZnO system, the position of the cantilever arm is detected by microvolt changes in the piezoresistive sensor's bridge circuit. This voltage is amplified to tens of volts to generate a suitable signal for feedback and analysis. Since the voltage gain is high, special considerations need to be given to electronic noise and phase delays in order to create a high speed, sensitive, electronic system. Automation is achieved through computer control of the microscope's operational parameters: force setpoint, gain, and feedback. A schematic of the automated microscope is shown in Figure 6.1.3. Two computers are used for imaging. One contains the custom electronic PC expansion board and is responsible for monitoring and modifying the cantilevers operational parameters. The other computer is responsible for collecting data and controlling the scanner.

The transfer function of the electronic system is shown in Figure 6.1.4. Figure 6.1.4a is a simulation of the combined electronic and mechanical system. The cantilever was modeled as a second order system with a resonance of 50 kHz and a Q of 100. These parameters were determined from the open loop transfer function of the cantilever when the tip is in contact with a sample (Figure 6.1.4c). The electronic system was

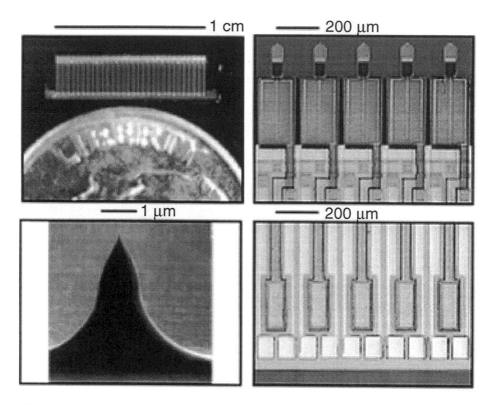

Figure 6.1.2 Array of cantilevers with integrated actuators and sensors with improved shielding between the actuator and sensor. Upper left: entire array of 50 cantilevers spanning 1 cm next to a dime for scale. Upper right: detail of five of the cantilevers spaced by 200 μm. Lower left: SEM micrograph of one of the single-crystal, silicon tips with a radius of curvature less than 10 nm. Lower right: corresponding electrical contact structure for the cantilevers. There are three leads per device: piezoresistor, ZnO, and tip bias.

modeled using the written specifications of each of the circuit components. Figure 6.1.4b shows the measured closed loop response of the cantilever and electronic system. We found excellent agreement between the simulations and the measured data. The primary limitation on bandwidth is due to the resonance of the cantilever. The influence of the electronic phase can be seen in the closed loop

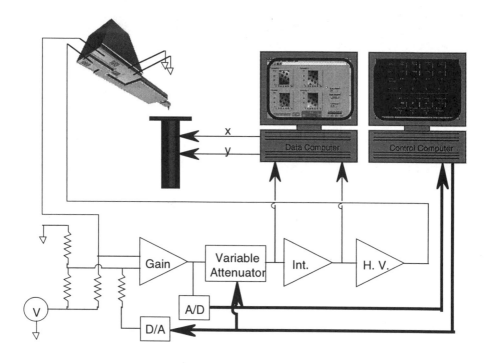

Figure 6.1.3   Schematic illustration of the automated, parallel, high-speed AFM. One computer controls the operation parameters of each cantilever while the other one collects data and controls the scan.

response in the small hump before the peak in the amplitude trace. The bandwidth of the amplifier chain is measured to be 1.1 MHz and the phase shift at 100 kHz is less than 10 degrees.

A 4x1 image obtained with the cantilevers and control electronics is shown in Figure 6.1.5. A detail of a memory cell on an integrated circuit chip is shown in Figure 6.1.5a. The scan size is just over 100μm x 100μm which is the limit of a conventional piezo tube. The ZnO actuator is able to track the 2μm high topography at 1mm/s. The images do not overlap because the tips are 200μm apart while the scan is

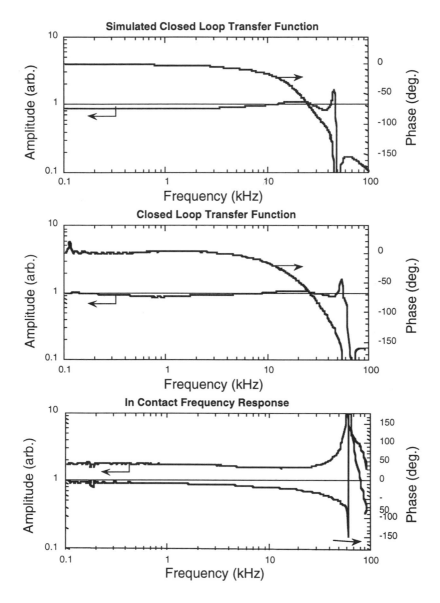

Figure 6.1.4   Transfer functions: (a) simulated closed-loop (b) measured closed-loop (cantilever plus electronics) (c) measured open-loop (cantilever only).

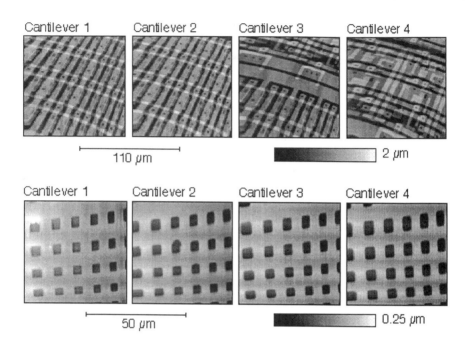

Figure 6.1.5 4 x 1 images of a memory cell (a) and a two-dimensional grating (b). The vertical sensitivity is less than 5 nm in a 20 kHz bandwidth. Tip velocity 1 mm/s (a) and 4 mm/s (b).

only 100 μm. Figure 6.1.5b shows a parallel image acquired at a scan speed of 4mm/s. The minimum detectable deflection for this image is less than 5nm in a 20kHz bandwidth.

6.2 Centimeter scale AFM[8]

The search for non-optical lithographic systems used to pattern 0.1um features is a daunting task, but as Alan Fowler points out in his Physics Today article on the subject, "If there is no attempt to find alternatives, they will never be found."[9] An array of scanning probes is one attempt to find the alternative for future systems.

The scanning probe is a tool for imaging and writing. The tips have been used to image atomic sized pixels, and to write sub-0.1um lines in positive and negative resists. But as it stands now, for wafer scale operation, the throughput is unacceptable. It is clear that the single tip scanning at modest speeds must be replaced with arrays of tips scanning at high speeds. Each tip in the array must operate independently to read, or write, individual pixels on demand.

Device oriented applications of scanning probe lithography are being pursued in many laboratories[10,11,12,13]. Additionally, resist systems that permit fast scanning[14], and parallelism[15] are being developed to address lithography throughput issues.

In this section we describe recent progress in imaging and performing lithography with cantilever arrays. The surface areas for the imaging and lithography data presented here are orders of magnitude larger than that of a typical atomic force microscope. This improvement is important since the areas imaged and patterned are commensurate with the areas of typical integrated circuit chips (100 mm^2).

---

8. Part of this section is reprinted with permission from Minne *et al.*, "Centimeter scale atomic force microscope imaging and lithography," *Applied Physics Letters* **73**, 1742 (1998). Copyright 1998 American Institute of Physics.

9. Alan Fowler, Physics Today, 50, 50 (1997).

10. E. S. Snow, P. M. Campbell, Science 270, 1639 (1995).

11. K. Matsumoto, M. Ishii, J. Shirakashi, B. J. Vartanian, and J. S. Harris, Proceedings of Quantum Devices and Circuits Alexandria, Egypt 4-7 June (1996).

12. H. T. Soh, K. Wilder, A. Atalar, C.F. Quate, Proc. 1997 Symposium on VLSI Technology, p. 129-130 (1997).

13. S. C. Minne, H. T. Soh, Ph. Flueckiger, and C. F. Quate, *Appl. Phys. Lett.* **67**, 703 (1995).

14. S. W. Park, H. T. Soh, C. F. Quate, S. –I. Park, Appl. Phys. Lett. **67**, 2415 (1995).

15. S.C. Minne, S.R. Manalis, A. Atalar, and C.F. Quate, J. Vac. Sci. Technol. B 14, 2456 (1996).

One-centimeter arrays (50 cantilevers on a 200 um period) are microfabricated with integrated piezoresistive sensors and integrated zinc-oxide (ZnO) actuators. The piezoresistive sensors provide 35Å resolution in a 20kHz bandwidth. More importantly, the integrated sensor simplifies the operation of the array because it requires no external components or alignment.

Figure 6.2.1 displays a 2mm x 2mm atomic force microscope (AFM) image of a memory cell on an integrated circuit taken with 10 cantilevers operating in parallel. Each of the 10 cantilevers swept out a vertical swath 200um x 2mm, with a pixel density of 512 x 5120. This corresponds to a pixel size of roughly 0.4 um. The pixel density of this image is not a fundamental limitation of the system or the data acquisition bandwidth, but arises due to technical limitations in processing the large amounts of data with computers available to us. The raw composite image size is over 100Mbytes. The inset of Figure 6.2.1 shows details contained within the acquired data that are not visible in the larger presentation. It should be noted that this area was not re-scanned, but rather the data was simply extracted from the larger file.

The inset of Figure 6.2.1 also spans the seam between two adjacent tips' imaging paths. By scanning over a distance greater than array period (200um), the adjacent tip's images can be stitched together to form a fully filled image. Conventional piezo-tubes are not suitable for this type of scanning due to their z-axis coupling. To solve this problem a custom flexure scanner (Nikon) was mounted on a long-range high-resolution scanner (Newport PM-500). The image was acquired at 1 mm/s (2.5 Hz over 200 um).

Figure 6.2.2 shows an image that spans 6.4 mm and was taken with 32 cantilevers operating in parallel. Again, since the scan distance exceeded the tip period, a full fill representation of the surface is obtained. The sample is a two dimensional grating with a period of 20 um and a step height of 200nm. The image has been broken into four strips and offset vertically for display purposes. The cantilevers were controlled by two custom built circuit boards interfaced to a PC. Each board contains the circuitry necessary to control 16 cantilevers. We believe this represents the largest parallel probe imaging operation to date. The small box in the corner of Figure 6.2.2 represents the maximum scan of a typical AFM.

High resolution imaging of large areas is useful in many applications. Furthermore, probe based imaging represents a starting point for probe based lithography. The imaging capabilities in a lithography system are useful for inspection of fabricated patterns, but more importantly, imaging capabilities represent the ability to lithographically overlay preexisting patterns.

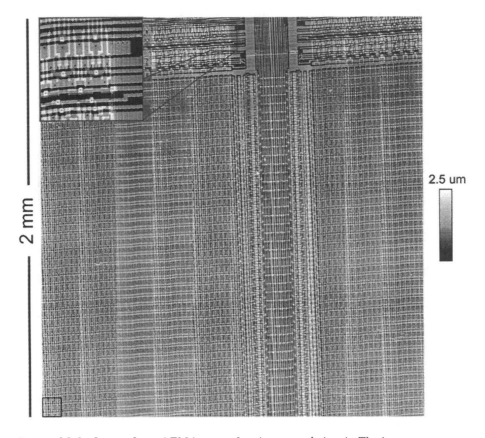

Figure 6.2.1   2mm x 2mm AFM image of an integrated circuit. The image was acquired in about 30 minutes with 10 cantilevers operating in parallel. The box at bottom left represents the maximum 100 μm x 100 μm scan from a conventional AFM.

In conclusion, the results presented here represent scan areas that are orders of magnitude larger than that of a typical atomic force microscope (0.01 mm2).These results are an essential step to covering even larger scan areas with increased throughput for both imaging and lithography. Work is currently underway to expand the size and functionality of the cantilever arrays, and the associated mechanical and electrical hardware, in order to operate in feedback mode over larger areas.

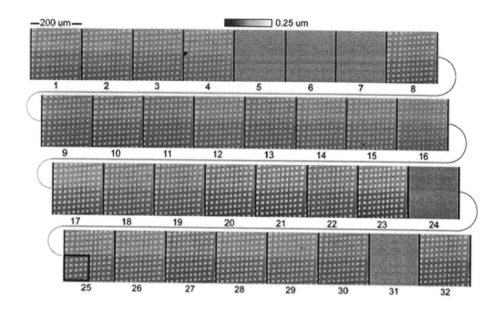

Figure 6.2.2 A 32 x 1 parallel AFM image of a two-dimensional diffraction grating. The horizontal distance is 6.4 mm and the entire image area is 1.28 mm².

CHAPTER 7

Scanning Probes for Information Storage and Retrieval

7.1 Introduction[1]

The role of the scanning probes for digital storage has yet to be defined, but we can discuss some significant steps that have been taken in this field. We will describe the work on writing bits on magnetic media, although there are other strategies for storing data with scanning probes. Mamin[2] has worked out a system where he creates small indents in PMMA for storing a single bit. Barrett[3] has employed the probe to store and read bits as trapped charge in films of silicon nitride. At Canon, Takimoto[4] has used the STM to store bits by changing the conductivity of the medium with a current pulse on the tip. Hosoka[5] has used small gold clusters deposited by field desorption from the probe tip.

1. Part of this chapter is reprinted with permission from Manalis et al., "Submicron studies of recording media using thin-film magnetic scanning probes," *Applied Physics Letters* **66**, 2585-2587 (1995). Copyright 1995 American Institute of Physics.

2. H. J. Mamin and D. Rugar, "Thermomechanical writing with an atomic force microscope tip," *Appl. Phys. Lett.* **61**, 1003-1005 (1992)

3. R. C. Barrett and C. F. Quate, "Charge storage in a nitride-oxide-silicon medium by scanning capacitance microscopy," *J. Appl. Phys.* **70**, 2725-2733 (1991).

4. K. Takimoto, H. Kawade, E. Kishi, K. Yano, K. Sakai, K. Hatanaka, K. Eguchi, and T. Nakagiri, "Switching and memory phenomena in Langmuir-Blodgett films with scanning tunneling microscope," *Appl. Phys. Lett.* **61**, 3032-3034 (1992)

Magnetic force microscopy (MFM) has emerged as a versatile tool for imaging surface magnetic fields of recording media and other materials down to the 10 nm scale. MFM works by scanning a tiny ferromagnetic probe over a sample and mapping its response to the sample's stray fields.[6,7] In addition to imaging, previous work has shown that solid magnetic probes can be used to write and sense 0.4 - 1 μm "bits" of reversed magnetization in recording media.[8,9,10]

In this chapter we show that scanning probes coated with a thin film of magnetic alloy[11] can produce magnetic bits down to 150 nm in perpendicular and magneto-optical (MO) recording media. Coated tips are easily batch-fabricated, and have more localized stray fields[12] than the solid tips[9,10] used in past work. Bits are written by bringing the tip into contact with the media and momentarily adding a uniform external field H_{ext} to the tip stray field. When the net field overcomes the local, or "point", coercivity, magnetization is reversed locally, and a bit is written. High-resolution imaging is done immediately with the same tip without perturbing the written pattern.

Using the ability to write and image bits, we have developed a new method for characterizing recording media coercivity on a sub-micron scale. Such measurements give crucial information about spatial media variations which lead to bit irregularities and increased data error rates.[13] By attempting to write many bits at each of several values of H_{ext}, a smooth, media-dependent write probability $P_w(H_{ext})$ can be measured. Semi-automation using a commercial magnetic force microscope allows rapid and convenient statistical analysis. For the media we examined, bits were written consistently ($P_w=1$) for sufficiently strong H_{ext}, none were written ($P_w=0$) for sufficiently weak fields, and writing was intermittent ($0<P_w<1$) for an intermediate field

5. S. Hosaka, A. Kikukura, H. Koyanagi, T. Shintani, M. Miyamoto, K. Nakamura, and K. Etoh, "SPM-based data storage for ultrahigh density recording," *Nanotechnology* **8**, A58-A62 (1997).

6. P. Grütter, H.J. Mamin, and D. Rugar, Scanning Probe Microscopy II (1991); P. Grütter, *MSA Bulletin* **24**, 416 (1994)

7. D. Rugar, H.J. Mamin, P. Guethner, S.E. Lambert, J.E. Stern, I. McFadyen, and T. Yogi, *J. Appl. Phys.* **68**, 1169 (1990)

8. J. Moreland and P. Rice, *Appl. Phys. Lett.* **57**, 310-312 (1990)

9. T. Ohkubo, J. Kishigami, K. Yanagisawa, and R. Kaneko, *IEEE Transactions on Magnetics* **6**, 5286 (1991); and *IEEE Trans. J. on Mag. in Jap.*, **8**, 245 (1993)

10. T. Goddenhenrich, U. Hartmann, and C. Heiden, *Ultramicroscopy* **42**, 256 (1992)

11. K. Babcock, M. Dugas, V. Elings, and S. Loper, *IEEE Transactions on Magnetics*, to appear.

12. P. Bryant, S. Schultz, and D.R. Fredkin, J. Appl. Phys. 69, 5877 (1991)

13. C.-J. Lin, J.C. Suits, and R.H. Geiss, *J. Appl. Phys.* **63**, 3825 (1988)

range. Intermittency is caused by spatial variations in the minimum localized field required to write a stable bit. These variations likely arise from spatial fluctuations in media structure or composition thought to govern domain nucleation and wall motion coercivity.[14,15]

Our results contrast with previous work writing bits (Ref. 9) which reported a dependence of bit strength and size with field, but no intermittent writing. In that work, the tips were solid Permalloy, and were less sharp than the probes used here. The stray fields of the Permalloy tips were likely too strong and insufficiently localized to probe the small-scale media fluctuations thought to occur on a scale <100 nm.[9,10] For media characterization, it appears that extremely sharp probes with thin-film coatings are essential.

The work described in this chapter was performed at Digital Instruments, Santa Barbara, CA, by S. Manalis, K. Babcock, J. Massie, and V. Elings, with M. Dugas of the Advanced Research Corporation, Minneapolis, MN.

7.2 Submicron recording with thin-film magnetic scanning probes

We used a NanoScope III (Digital Instruments) scanning probe system with a Multi-Mode[16] microscope and a piezoelectric scanner with a 120 μm lateral range. The probes were single-crystal silicon cantilevers with pyramidal tips sputtered with a 400 Å Co-Cr alloy.[11] The tip was magnetized with its stray field opposing the media magnetization. Magnetic writing was accomplished in contact mode, with laser beam deflection used to detect the cantilever position. Digital feedback controlled the vertical scanner position so as to keep the cantilever deflection, and hence tip-media force, constant during the write process; see Figure 7.2.1. To write a bit, a uniform external field H_{ext} was momentarily added in the direction of the tip stray field. A 1.8 cm diameter coil mounted on the scanner produced external field pulses up to 3000 Oe with a 280 μs rise time. Current was supplied by an amplifier driven by a software-controlled analog signal from the microscope controller. Lateral tip motion and exter-

14. M. Mansipur, R. Giles, and G. Patterson, *J. Appl. Phys.* **69**, 4844 (1991)

15. K. O'Grady, T. Thomson, J.J. Greaves, and G. Bayreuther, *J. Appl. Phys.* **75**, 6849 (1994)

16. MultiMode, TappingMode, and LiftMode are trademarks of Digital Instruments. Tapping-Mode and LiftMode, V. Elings and J. Gurley, U.S. Patent Nos. 5,266,801 and 5,308,974, Digital Instruments, Santa Barbara, CA.

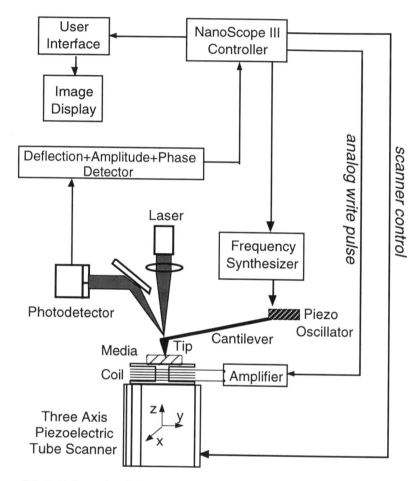

Figure 7.2.1 Schematic of the thin-film, magnetic scanning probe set-up.

nal field pulse strength and duration were controlled with the Nanoscope III lithography software. Typical test patterns were produced by repeating a 1 ms field write pulse, 1 s wait, and 1 μm translation sequence.

Using the same tip, the magnetic bit pattern and the surface topography were immediately imaged by oscillating the cantilever at its resonant frequency while scanning the media in a raster pattern. A magnetic image was produced by mapping phase shifts in the cantilever oscillation caused by gradients in the magnetic force exerted on the tip by the sample's stray fields.[1,2] Figure 7.2.2 shows a representative magnetic force

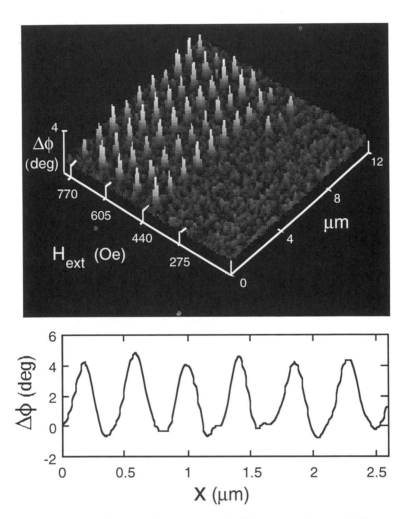

*Figure 7.2.2   Magnetic force gradient image of a bit array written on bilayer
perpendicular media. Lower trace: cross-section of a higher density
array of 180 nm bits spaced 370 nm apart.*

gradient image and section plot of bits written on perpendicular media consisting of a
NiFe sublayer and a CoCr top layer. Simultaneous imaging of topography and magnetic field gradients were done in two passes across each raster scan line. On the first
pass, the topography was recorded by lightly tapping the surface (TappingMode[16]).

LiftMode[16] was used on the second pass to raise the tip to a selected height above the local surface topography (typically 25-50 nm). The tip-surface separation is held constant, and cantilever oscillations are sensitive to magnetic force gradients without being influenced by topographical features. Because of the relatively small stray field produced by thin-film probes,[11,12] this imaging procedure does not affect the written bit pattern, even at the low lift heights (<50 nm) that give high resolution. This contrasts with past work using solid tips[4,5] which required relatively large tip-sample separations for non-destructive imaging.

Also shown in Figure 7.2.2 is a cross section of bits written on the same perpendicular media at a higher density than the pictured array. The bits are 180 nm wide (half maximum), with 400 nm spacing, comparable to the background magnetic fluctuations of ~150 nm. A given tip writes similar size bits (typically 150-300 nm) on both perpendicular media and amorphous TbGdFeCo MO media (Figure 7.2.3) having smaller-scale background fluctuations. This suggests that bit size is primarily determined by the rapid drop-off in stray field away from the tip.[12] Recording and imaging were also possible with Permalloy ($Ni_{81}Fe_{19}$) coated tips. However, the bit sizes were less consistent than those written with Co-Cr coated tips, possibly because the low-coercivity Permalloy is altered unpredictably by the changing stray fields of the media during the write process.

Point coercivities for various recording media were evaluated by writing rows of bits, using a successively weaker H_{ext} for each row. In the bit array of Figure 7.2.2, the field H_{ext} was pulsed ten times for each row at positions spaced 1 μm apart, and decremented 85 Oe in successive rows. For sufficiently large H_{ext}, bits were written consistently; all ten bits were written in a row for $H_{ext} \geq 440$ Oe. Using the same tip, similar results were obtained for TbFeCo MO media having bulk coercivity[17] Hc=1350 Oe, except that bits were written consistently for $H_{ext} \geq 600$ Oe. Fields ~160 Oe weaker than these thresholds resulted in no bits being written. Note that the write threshold is lower in the perpendicular media than in the MO, even though the former has coercivity[13] Hc=2300 Oe, *higher* than the MO value.

For all the media examined, we found an intermediate field range for which the writing of bits is intermittent. Figure 7.2.3 shows a bit test pattern written on TbGdFeCo MO media with bulk coercivity12 Hc=4000 Oe. Again, H_{ext} was pulsed ten times with a 1 μm spacing within each row, and decremented 170 Oe between rows. For this particular array, bits were written intermittently over an external field range $1190 < H_{ext} < 2365$ Oe. We also found an intermittent field range for the lower-coerciv-

---

17. Measured via vibrating sample magnetometry; 3M Center, St. Paul, MN.

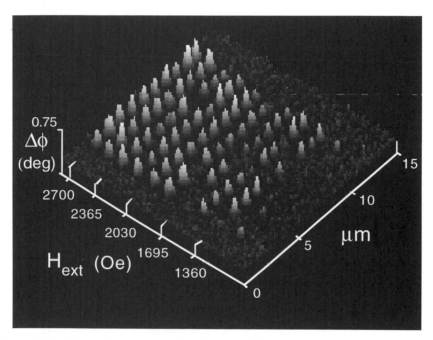

*Figure 7.2.3   Magnetic force gradient imaged of bits on H_c=4.0 kOe TbGdFeCo
MO media, with H_{ext} decremented 170 Oe between rows. Bits were
written intermittently for 1190<H_{ext}<2365 Oe.*

ity media described above when sufficiently small field decrements were used. The
intermittent range was roughly 80 Oe for the perpendicular media, and 270 Oe for the
Hc=1350 Oe MO media.

The intermittent range can be characterized by a write probability $P_w(H_{ext})$. We
obtained good statistics by attempting to write large numbers of bits at each value of
H_{ext}. Figure 7.2.4a compares Hc=1350 Oe MO with the perpendicular media, and
Figure 7.2.4b compares Hc=1350 Oe with Hc=4000 Oe MO media. For each field
value H_{ext}, 20 field pulses were applied in a row, and the write probability taken as
P_w=(number of bits written)/20. The results were consistent for data taken from three
different areas of each sample, showing that P_w is reproducible and that intermittency
is caused by a mechanism on a small length scale. The curves $P_w(H_{ext})$ explore the
distribution of regions having various point coercivities. For example, $P_w(H_{ext})$=0.6
indicates that 60% of the media surface has point coercivities below H_{ext}.

For comparison, the dashed curves in Figure 7.2.4 show scaled bulk hysteresis loops

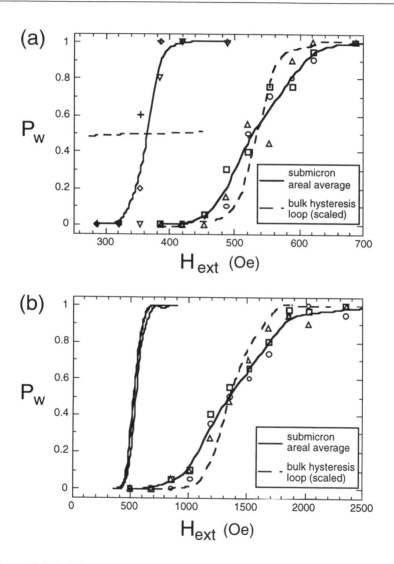

Figure 7.2.4   Write probability vs. external field.

taken of the same ~1/2 cm^2 samples used to obtain the P_w data. The raw data M(H) was scaled according to $\overline{M}(H) = a[M(H + \Delta H) + b]$, where the scale factor a and offset b bring the asymptotic magnetization values to match the limits $P_w=0,1$. The total field during the write process is uncertain because the tip stray field is not known. We chose the field offset ΔH to align the transition region of the bulk curves with the intermittent region of P_w. For the perpendicular media in Figure 7.2.4a, P_w is significantly sharper than the transition region of the hysteresis loop,[18] with a slope two orders of magnitude steeper at the transition. This suggests that the media is far more "square" with respect to actual writing than conventional measurements would indicate. In contrast, P_w is broader than the bulk hysteresis loops for both MO media. As shown in Figure 7.2.4b, the higher coercivity media (Hc=4000 Oe) has an intermittent range ($P_w\neq0$ or 1) of $700<H_{ext}<2500$ Oe, nearly 45% of the bulk coercivity. This is significanly broader than the transition region ($1000<H<1800$ Oe) of the bulk hysteresis loop.

The results for P_w were consistent from tip-to-tip, as demonstrated in Figure 7.2.4b by using three different tips to measure P_w on the perpendicular media. In general, we found that the form of P_w is very consistent between tips, with ~100 Oe variation in H_{ext}.[19] This indicates a high degree of uniformity in the magnitude and spatial distribution of the stray fields of the tips we used. Even without knowledge of the tip properties, different media can be compared quantitatively by using the same tip.

We also tried writing bit arrays with two 1 ms pulses at each array site, spaced by 20 ms, and compared the results to those obtained with one pulse at each site. At fields H_{ext} giving $P_w =1/2$, multiple attempts did not increase the write probability, confirming that intermittency is caused solely by spatial variations in the media, and is not characterized by a write probability per attempt. Arrays with 2 µm bit spacing gave the same P_w as the 1 µm arrays, verifying that the results were not affected by demagnetization fields at the edge of bit arrays.

In conclusion, we have demonstrated a technique for probing the coercivity of perpendicularly-magnetized recording media on a 100 nm scale using a commercial MFM and an electromagnet. Writing and immediate imaging of bits provides a direct assessment of recording performance, with a spatial resolution surpassing other local probes based on the Kerr effect.[20,21] Typical arrays were written at rate of ~0.5 bits per second using the lithography software of the Nanoscope III controller and imaged

---

18. Measured via polar Kerr effect; Censtor Corp., San Jose, CA. This technique averages the response of the surface layer over an area ~1 mm^2.

19. This is consistent with previous work (Ref. 6) which found a 15% variation in imaging sensitivity between tips.

20. H.-P.D. Shieh and M.H. Kryder, IEEE Trans. Mag. 24, 2464 (1988).

---

in a few minutes, giving rapid results for different media. The write probability $P_w(H_{ext})$ characterizes intermittent bit writing due to spatial media fluctuations, and complements bulk hysteresis measurements done in the presence of a uniform applied field. The difference was dramatic in the case of bi-layer perpendicular media, with a P_w transition two orders of magnitude sharper than the hysteresis loop transition. The MO media showed the opposite trend, with P_w being significantly broader than the "square" bulk hysteresis loops.

Point coercivity measurements may give insight into micromagnetic mechanisms which govern media performance. Intermittency in writing may be due to variations in nucleation fields or vestigial reversed domains, perhaps related to composition fluctuations or grain structure. The sigmoidal shape of the P_w curves shown here suggest a Gaussian distribution of nucleation fields. We have also found that sufficiently large field pulses can cause existing bits to grow. Subsequent imaging may allow domain wall coercivities, and the time dependence of reversal reported in multilayers,[10] to be investigated directly.

21. C. D. Wright, W.W. Clegg, A. Boudjemline, and N. Heyes, Jpn. J. Appl. Phys. 33, 2058 (1994)

CHAPTER 8

Silicon Process Flow: ZnO actuator and piezoresistive sensor

8.1 Introduction

This chapter provides a step by step process for fabricating the ZnO cantilever. We begine by determining values for the piezoresistive coefficient for different doping levels. This is most easily done using the tables published by Kanda[1]. Figure 8.1.1a is the plot of piezoresistive coefficients for the different orientations of silicon. We are primarily concerned with (110) orientation in the top half of the figure, which gives the longitudinal coefficient, when the current and stress are in the same direction. Figure 8.1.1b gives a modifier, which depends on doping and temperature, to the result in found in Figure 8.1.1a.

8.2 Process Flow

The fabrication of a piezoresistive cantilever with a piezoelectric actuator is a ten mask process. The cantilever is fabricated on an SOI wafer which is etched from the back so the cantilever will be freestanding when the processing is complete. On the

---

1. Y. Kanda, "A Graphical Representation of the Piezoresistance Coefficients in Silicon," IEEE Trans. Ele. Dev., **ED-29**, 64 (1982)

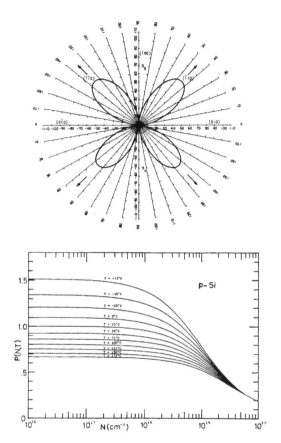

Figure 8.1.1 *Top: piezoresistive coefficient for different silicon orientations.*
Bottom: dependence of the piezoresistive coefficient on temperature
and doping. From Kanda Ref. 1.

cantilever a tip, a piezoresistor, and a zinc-oxide (ZnO) actuator are fabricated. The
tip is made at the far end of the cantilever by a undercutting a silicon mask. The
piezoresistor is implanted in the middle of the cantilever, and the ZnO actuator is
grown at the base. The final device is shown in Figure 8.2.1. The following is a
detailed process flow; the "steps" refer to the diagrams located at the end of this section.

Figure 8.2.1   Finished device

Step 0: The cantilevers are fabricated on a silicon-on-insulator (SOI) <100> wafer. The SOI wafer consists of a one micron layer of oxide sandwiched between a 450 micron layer of p-type <100> single crystal silicon, and a 10 micron layer of intrinsic <100> single crystal silicon.

Step 1-3: A 1 micron layer of thermal oxide is grown on the wafer. A two sided lithography and an oxide etch define the tip mask and the back-side etch windows. The oxide thickness was chosen such that further processing will not obscure the back-side windows, which are not used until the last step.

Step 4-5: The top surface of the wafer is etched in a dry silicon etch. The etch is allowed to continue until the tip mask is completely undercut, leaving a sharp tip. The top silicon layer will eventually form both the tip and the cantilever. The cantilever should be between 3 and 4 microns thick leaving 6 to 7 microns of silicon to be etched for the tip. It is necessary to experiment with the top mask size and the plasma etcher to determine the configuration that will obtain a sharp tip after a 6 to 7 micron silicon etch.

Step 6: Etch the oxide from the front surface of the wafer. Oxidation sharpen the tip[2] by growing and removing thermal oxide. Oxidation sharpening works because the oxidation rate of convex silicon corners is slower than that of flat silicon. The thermal oxidation will also grow oxide on the back of the wafer. The large thickness of the initial oxide will keep the original back-side pattern defined.

Step 7-9: Define the piezoresistor by implanting 5e14 cm2 boron at 80keV over the entire wafer. Perform the second implant photolithography and implant the base of the cantilevers with 5×10^{16} cm2 boron at 80keV. SUPREM IV simulations for the implants are shown in Figure 8.2.2.

The two implants are necessary to keep the ZnO actuator from interfering with the piezoresistor. The second implant dopes the resistor under the actuator roughly 100 times more than the piezoresistor. This additional doping will make the contribution of the resistors under the ZnO actuator to the piezoresistive effect negligible (see section 2.7 and 2.8).

Step 10: Grow a thin oxide to protect and insulate the cantilever and tip. An interesting phenomenon called doping enhanced oxidation occurs during this process. Recall that the area under the cantilever is heavily doped with boron to 10^{21} cm^{-3}. When boron doped silicon oxidizes, the boron within the silicon becomes incorporated into the newly grown oxide. The incorporated boron weakens the "glassy", or bond, structure of the oxide, allowing H_2O and O_2 to diffuse through it more quickly.[3] This effect only becomes apparent at boron densities greater than 10^{20} cm^{-3}. Fortunately, we can simulate this and design our device such that the oxide nonuniformities do not interfere with our processing. The effect of doping enhanced oxidation of silicon can clearly be seen in the oxide thicknesses variations in figure Figure 8.2.2

2. D. Kao, J. McVittie, W. Nix, K. Saraswat, "Two-dimensional thermal oxidation of silicon. II. Modeling stress effects in wet oxides." IEEE Trans. Electron Devices Vol. ED-34, no 1, pp1008, 1987.

3. S. Wolf and R. N. Tauber, "Silicon Processing for the VLSI Era,", Vol 1, pp 213, Lattice Press, Sunset Beach, California (1986).

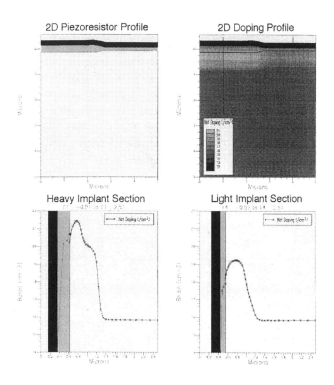

2D Piezoresistor Profile   2D Doping Profile

Heavy Implant Section   Light Implant Section

*Figure 8.2.2   SUPREM doping simulations of one and two dimensional boron
implants in the lightly and heavily doped regions of the cantilever.*

Step 11-12: Perform the cantilever lithography. Etch through the thin oxide of step 10.
Completely etch the 3 - 4 µm cantilever silicon. The middle oxide of the SOI structure
should now be exposed.

Step 13 - 14: Grow 0.2 microns low stress PECVD nitride. Perform nitride photolithography, and RIE etch the nitride. The nitride acts as a buffer layer between the cantilever and the ZnO actuator. Step 25 etches away the middle oxide from the back. If
the nitride buffer layer were not present, the back-side etch could attack the cantilever
insulation or the ZnO actuator. Low stress PECVD nitride is used to minimize the
stress in the cantilever. Any stress in the finished cantilever will manifest itself as
undesirable curling.

Step 15 - 17: Perform the contact lithography and etch the contact holes. Evaporate, pattern and etch 0.5 microns of gold. Strip the metal photoresist in an oxygen plasma. Metal 1 makes contact to the piezoresistors and defines the bottom contact for the ZnO actuator.

Step 18: Plasma clean the surface. Deposit 3.5 microns of oriented ZnO [4]. The plasma clean removes contaminants from the surface of the Gold. The cleanliness of the surface will determine how well oriented the ZnO layer grows.

Step 19 - 20: Perform the ZnO lithography. Etch the ZnO in 15 g NaNO3 + 5 ml HNO3 + 600 ml H2O [5]. Strip photoresist in acetone and an oxygen plasma. This ZnO etch has excellent selectivity against nitride and oxide, and does not attack gold. Note: the diagrams at the end of this section are not to scale. The ZnO sandwich is 5 microns above the surface of the cantilever and the tip is 6 - 7 microns above the surface of the cantilever.

Step 21: Perform the top electrode photolithography. Soak wafer in chlorobenzene for five minutes (optional). Evaporate 0.5 microns of gold for lift-off. Lift-off lithography is used because most gold etches attack ZnO.

Step 22: All of the gold evaporations have been 0.5 microns. This may not be enough gold for wire bonding. An extra lift-off lithography can be used to increase the thickness of the gold on the bond pads.

Step 23 - 26: Spin coat the top surface with polyimide and cure. Etch the backside window until the silicon is completely exposed. Etch through the bulk of the silicon in an EDP etch. Etch out the middle oxide, and remove the polyimide in a high power oxygen plasma.

The fabrication is now complete. The wafer must be cleaved and the individual cantilevers mounted and wire bonded before use. Figure 8.2.3 shows a labeled perspective view of the finished device.

4. J. C. Zesch, B. Hadimioglu, B. T. Khuri-Yakub, M. Lim, R. Lujan, J. Ho, S. Akamine, D. Steinmetz, C. F. Quate, and E. G. Rawson, "Deposition of highly oriented low-stress ZnO films" UltraSonics Symposium Proceedings, B. R. McAvoy, New York, N.Y., **1**, pp445, (1991).

5. T. Albrecht "Advances in atomic force microscopy and scanning tunneling microscopy" PhD. Thesis, Stanford University, 1989.

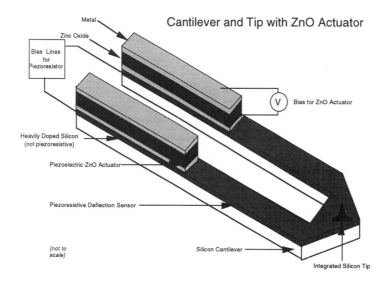

Metal

Zinc Oxide

Cantilever and Tip with ZnO Actuator

Bias Lines
for
Piezoresistor

V   Bias for ZnO Actuator

Heavily Doped Silicon
(not piezoresistive)

Piezoelectric ZnO Actuator

Piezoresistive Deflection Sensor

(not to scale)

Silicon Cantilever

Integrated Silicon Tip

Figure 8.2.3   Schematic diagram of the finished device

TABLE 2. Cantilever process flow

0. Starting Material: 4", <100> SOI	Top Si = 10 microns
	Middle Oxide = 1 micron
	Bottom Si = 450 microns
1. Grow 1micron of oxide	Oxidize Wafer
2. Tip/Backside photolithography	

TABLE 2. Cantilever process flow

3. Etch Oxide	Pattern Tip and Backside
4. Dry Silicon Etch	Undercut oxide cap in dry etch to produce a tip.
5. Strip Photoresist	

TABLE 2. Cantilever process flow

6. Oxidation Sharpening 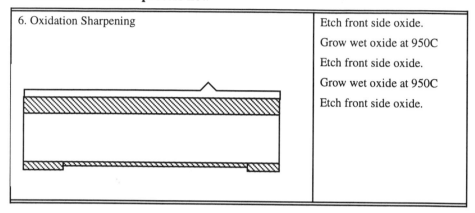	Etch front side oxide. Grow wet oxide at 950C Etch front side oxide. Grow wet oxide at 950C Etch front side oxide.

TABLE 2. Cantilever process flow

7. Implant piezoresistor	Dose = 5e14 cm^2
	Energy = 80keV
	Boron
8. Define and Implant Conductor	Implant2 photolithography
	Dose = 5e16 cm^2
	Energy = 80keV
	Boron
9. Strip Photoresist	

TABLE 2. Cantilever process flow

10. Grow Thin Oxide	Grow thin oxide.
11. Define Cantilever	Cantilever Photolithography. Etch Oxide Etch Silicon
12. Strip Photoresist	

TABLE 2. Cantilever process flow

13. Buffer Nitride	Grow 0.2 microns of low stress PECVD nitride.
14. Pattern and Etch Nitride	Pattern and Etch Nitride
15. Etch Contacts	Pattern and Etch Contacts

TABLE 2. Cantilever process flow

16. Metal 1	Evaporate 0.5 micron Au.
17. Metal 1 Photolithography	Pattern and Etch Gold
18. Deposit ZnO Sandwich	Deposit 3.5 microns ZnO

TABLE 2. Cantilever process flow

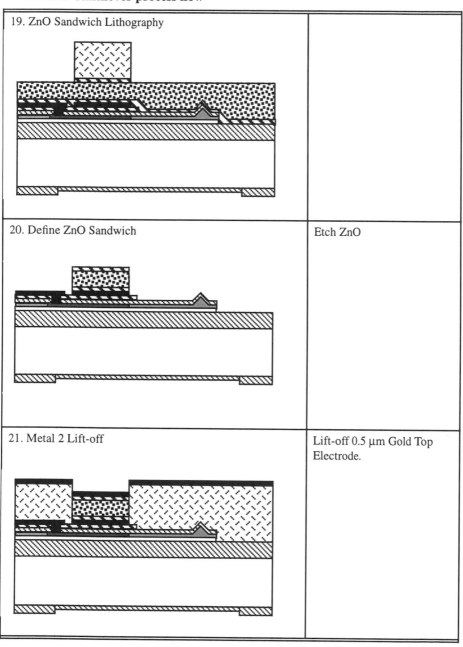

19. ZnO Sandwich Lithography	
20. Define ZnO Sandwich	Etch ZnO
21. Metal 2 Lift-off	Lift-off 0.5 μm Gold Top Electrode.

TABLE 2. Cantilever process flow

22. Re-metallize Bonding Contacts 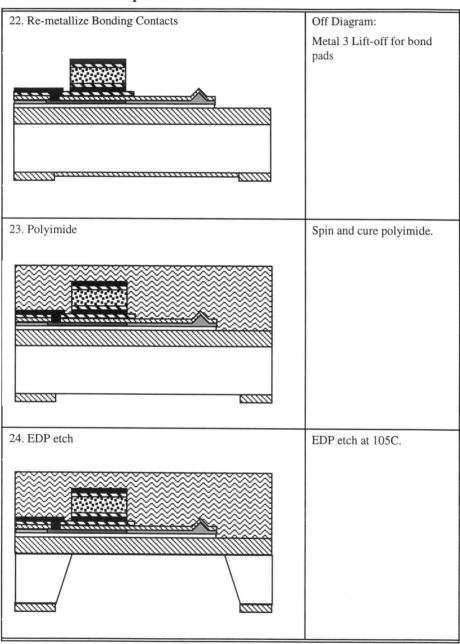	Off Diagram: Metal 3 Lift-off for bond pads
23. Polyimide	Spin and cure polyimide.
24. EDP etch	EDP etch at 105C.

TABLE 2. Cantilever process flow

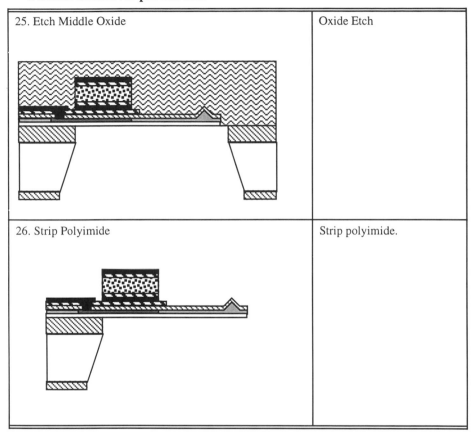

25. Etch Middle Oxide	Oxide Etch
26. Strip Polyimide	Strip polyimide.

CHAPTER 9

Silicon Process Flow: Interdigital Cantilever

9.1 Fabrication Process

The interdigital cantilever is fabricated using a standard cantilever process. We started with a <100> SOI wafer on which the top silicon is 10 μm of undoped epitaxial silicon. We begin by growing one micron of thermal oxide on the wafers followed by LPCVD nitride deposition. Tip masks are patterned into the nitride with a plasma etch and then into the oxide with 6:1 HF. The tips are formed by undercutting the silicon with an isotropic plasma etch. The tip height is increased with an anisotropic etch until the desired cantilever thickness is reached. The tips are then sharpened by growing and removing thermal oxide. The oxidation sequence sharpens the silicon tip because the oxidation rate of convex silicon corners is slower than that of flat silicon. The result is a tip with an end radius of ~100 Å. The cantilever and the interdigitated fingers are defined in a plasma etch. The top surface is then passivated with polyimide and the bulk silicon is etched with ethylene diamine pyrocathecol (EDP) using the middle oxide as an etch stop. Cantilevers are released by etching the middle oxide in 6:1 HF and removing the polyimide in an oxygen plasma.

A scanning electron micrograph (SEM) of an interdigital cantilever that is 220 μm long, 150 μm wide, and 3 μm thick is shown in Figure 9.1.1. The interdigitated fingers are 3 μm wide and are defined in an anisotropic etch along with the cantilever and are shown in detail in .Figure 9.1.2. A zoom-in of the single crystal silicon tip is shown in Figure 9.1.3.

Figure 9.1.1   SEM micrograph of an interdigital cantilever that is 220 um long, 150 um wide, and 3 um thick.

Figure 9.1.2   Detail of 3 um wide interdigitated fingers

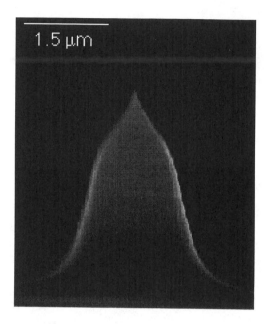

Figure 9.1.3   Zoom in on the tip

TABLE 3. Interdigital Cantilever Process Flow

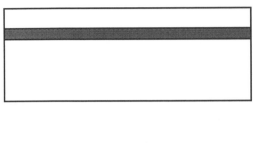

1. Silicon on insulator (SOI) wafer

Top Si: 10 μm

Middle Oxide: 1 μm

Bottom Si: 450 μm

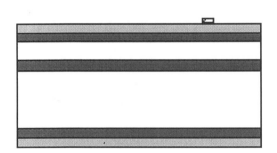

2. Grow 1 μm of oxide

-wet oxidation at 1100 C for 2 hrs 25 min

3. Grow 1 μm of nitride

4. Pattern TIP mask with photoresist.

TABLE 3. Interdigital Cantilever Process Flow

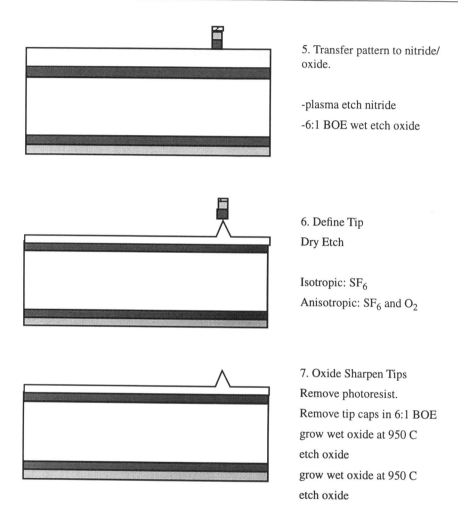

5. Transfer pattern to nitride/oxide.

-plasma etch nitride
-6:1 BOE wet etch oxide

6. Define Tip
Dry Etch

Isotropic: SF_6
Anisotropic: SF_6 and O_2

7. Oxide Sharpen Tips
Remove photoresist.
Remove tip caps in 6:1 BOE
grow wet oxide at 950 C
etch oxide
grow wet oxide at 950 C
etch oxide

TABLE 3. Interdigital Cantilever Process Flow

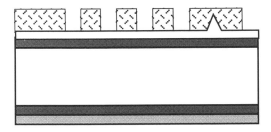

8. Pattern cantilever and inter-digitated fingers

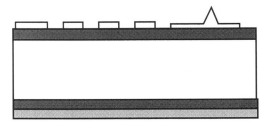

9. Etch silicon with dry aniso-tropic etch

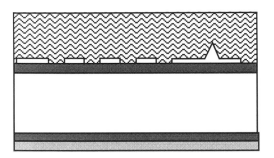

10. Protect frontside with thick layer of polyimide.

TABLE 3. Interdigital Cantilever Process Flow

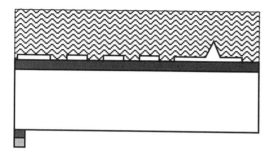

11. Pattern Backside Mask

Dry etch nitride

Wet etch oxide

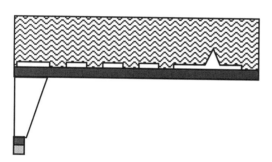

12. Remove substrate silicon

Wet etch bottom layer in hot EDP for 8-9 hours.

TABLE 3. **Interdigital Cantilever Process Flow**

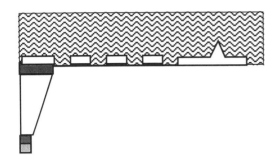

13. Etch middle oxide layer in 6:1 BOE.

14. Remove polyimide in oxygen plasma.